Climate Change and Food Security

ADVANCES IN GLOBAL CHANGE RESEARCH

VOLUME 37

For other titles published in this series, go to
www.springer.com/series/5588

David Lobell · Marshall Burke
Editors

Climate Change and Food Security

Adapting Agriculture to a Warmer World

Editors
David Lobell
Stanford University
CA, USA
dlobell@stanford.edu

Marshall Burke
Stanford University
CA, USA
mburke@stanford.edu

ISBN 978-90-481-2952-2 e-ISBN 978-90-481-2953-9
DOI 10.1007/978-90-481-2953-9
Springer Dordrecht Heidelberg London New York

Library of Congress Control Number: 2009928835

Printed on acid-free paper

Springer is part of Springer Science+Business Media (www.springer.com)

Contents

Part I

1 **Introduction** ... 3
David Lobell and Marshall Burke

2 **Climate Effects on Food Security: An Overview** ... 13
Marshall Burke and David Lobell

3 **Climate Models and Their Projections of Future Changes** ... 31
Claudia Tebaldi and Reto Knutti

Part II

4 **Crop Response to Climate: Ecophysiological Models** ... 59
Jeffrey W. White and Gerrit Hoogenboom

5 **Crop Responses to Climate: Time-Series Models** ... 85
David Lobell

6 **Crop Responses to Climate and Weather: Cross-Section and Panel Models** ... 99
Wolfram Schlenker

7 **Direct Effects of Rising Atmospheric Carbon Dioxide and Ozone on Crop Yields** ... 109
Elizabeth A. Ainsworth and Justin M. McGrath

Part III

8 **Food Security and Adaptation to Climate Change: What Do We Know?** ... 133
Marshall Burke and David Lobell

9 Breeding Strategies to Adapt Crops to a Changing Climate 155
R.M. Trethowan, M.A. Turner, and T.M. Chattha

Part IV

10 Global and Regional Assessments 177
David Lobell and Marshall Burke

11 Where Do We Go from Here? 193
David Lobell and Marshall Burke

Index 199

Part I

Chapter 1
Introduction

David Lobell and Marshall Burke

1.1 Why Read This Book?

The Earth is clearly warming. Mounting evidence from around the globe has removed virtually any serious doubt over this fact, and also over whether the main culprit is human emissions of heat-trapping gases such as carbon dioxide (CO_2). These emissions have been the product of a march towards better economic living standards, and for much of the world this march has led people out of a life of hunger and poverty and into one of relative comfort and security. But many have been left behind, and roughly 1 billion people continue to live under poverty and with insecure access to food. In an average day, more than 20,000 children die from hunger related causes.

A large majority of the world's poor continue to live in rural areas and depend on agriculture for their livelihoods. Given that agriculture everywhere remains dependent on weather, changes in climate have the potential to disproportionally affect these poor populations. But what, precisely, will human-induced climate changes mean for the globe's billion poor? How will climate change interact with the many other factors that affect the future of food production and food security?

There are no easy answers to these questions. That fact, of course, does not stop people from making simple predictions based on ideology, such as that innovation and free market responses will avoid any damages, or that climate change will wreak havoc on humans. Theory alone cannot refute either of these extreme positions, as there are no obvious reasons why that the pace of climate change caused by human activity should or should not match the pace with which we are able to adapt food production systems. Rather, the issue at hand is an empirical one, and finding answers will require a cadre of scientists capable of collecting and analyzing the relevant data, and policy makers and citizens capable of understanding their implications.

This book aims to foster these capabilities in students, researchers, and policy-makers in the field, by providing an accessible introduction into the fundamental

D. Lobell(✉) and M. Burke
Stanford University, CA, USA

D. Lobell and M. Burke (eds.), *Climate Change and Food Security*,
Advances in Global Change Research 37, DOI 10.1007/978-90-481-2953-9_1,

science needed to address the potential effects of climate change on food security. To accomplish this, our approach in the book has five main features that we feel distinguish it from other texts on this subject.

The first is the style of presentation: we offer non-technical descriptions of the fundamental data and science underlying models of agricultural impacts, descriptions that should be comprehensible without prior knowledge of the subject. We hope this will prove useful not only for students, but also for inter-disciplinary researchers and policy makers who wish to understand in more detail the output from models in disciplines with which they commonly interact. Accordingly, we have not attempted an exhaustive review of recent applications in any of the chapters – results that in any case are likely to change quickly – but rather present enough examples to explain important concepts.

The second feature is a focus on the full suite of interactions between climate change and food security, which moves beyond the traditional narrower focus on the potential climate effects on the production of a few cereal crops. Although the main cereal crops (rice, wheat, maize) do contribute the majority of calories consumed globally, in many parts of the developing world they play a more minor role, warranting increased attention to the myriad other crops of importance to the poor. Furthermore, the majority of poor households are both producers and consumers of agricultural commodities, suggesting that a narrow focus on production might miss the effects of climate change on other important aspects of food security, such as incomes and health.

A third related feature of the book is our focus on the inherent uncertainties associated with any assessment of the effects of climate on food security, uncertainties that are often not clearly quantified elsewhere in the literature. In each chapter we focus on the types of uncertainties that exist and the ways that researchers attempt to measure them. We feel this reflects a broader trend in the community to move away from simple "best guess" estimates and provide a more probabilistic view of the future.

The fourth feature of the book is an embrace of the diversity of approaches and perspectives necessary for a complete assessment of the linkages between climate and food security. Because a complete assessment requires the integration of tools from often diverse fields, the book presents a broad range of perspectives from various experts. Thus, for example, the presentation does not focus on a single approach to estimating crop responses to climate change but covers the strengths and weaknesses of the various methods employed by researchers on this particular topic.

Finally, the book discusses extensively the adaptation options available to agriculture in order to cope with climate change over the next few decades. This contrasts with many studies that have focused instead on the longer time frame of 2080 or 2100. In our opinion, this choice reflects a broader trend in the global change community to focus not only on questions of mitigation (i.e. whether and how to reduce greenhouse gas emissions) but increasingly also on how to adapt to the changes we can expect regardless of emissions reductions.

1.2 The Strength and Limits of Models

In order to provide quantitative measures of climate change impacts, we will rely heavily on numerical models of various pieces of the puzzle, including climate, agricultural, and economic systems (Fig. 1.1). Models are needed because it is rarely possible to perform controlled experiments where one or two factors are changed while others are held constant, particularly for the time scales and spatial scales of interest. One cannot measure, for example, global crop production with climate change and compare it to a world without. Instead, one must perform the controlled experiments in the simplified world of computer models, which can be run at any scale.

However, it is important to remember that models are only simplified representations of reality, tools that can be used to estimate things that often cannot be directly measured. When their output is compared to things that can be measured, they almost always contain some error. In the case of predicting the future, this error arises both from not knowing perfectly how the climate and agricultural systems currently behave, and not knowing the future decisions that humans will make (both on the mitigation and adaptation side) that will influence the result.

The goal of modeling must therefore be to estimate not only a "best-guess", but also a probability distribution function (pdf), which describes the probability that the true value will take on each possible value. Often of interest is the chance that a particular threshold will be exceeded, such as 500 ppm atmospheric CO_2, 2°C global average annual temperature, or 1 billion food insecure people. For these purposes, a single best guess of impacts is essentially useless. While nearly everyone acknowledges that treating a single output of a model as a firm "prediction" can be foolish, there appears a strong and persistent desire in humans to ignore uncertainties and overstate confidence in predictions.

Of course, the alternative of throwing up our hands and claiming no knowledge about the future is equally unattractive. Instead, we seek to clearly distinguish between those aspects of the future we know well and those that we do not – a task that can only be achieved by tracking uncertainties. The job is made somewhat easier by the fact that the goal is often not to actually predict the future, but instead to predict the difference between two outcomes. For example, impacts on wheat in China versus India; impacts on corn versus rice; impacts for low versus high CO_2 emissions; or impacts for low versus high investments in a certain adaptation technology. In these cases, errors that are similar for each individual projection will tend to cancel out when looking at differences. It is thus often helpful to remember that while we would love to be able to predict everything about the future, our actual goals (and certainly our abilities!) are often much more modest.

Fig. 1.1 The cascade of models needed to evaluate the impacts of climate change on food security

1.3 The Importance of Time Scales

Two main types of interventions are often discussed as ways to reduce the impacts of climate change on society: mitigation and adaptation. Mitigation is a reduction in greenhouse gas emissions, which leads (eventually) to a reduction in climate change. Adaptation refers to changes made to a system impacted by climate, in this case some aspect of the food economy, that improve the outcome of climate change relative to no adaptation. Adaptations can include both changes that either reduce negative outcomes or enhance positive outcomes.

The effectiveness of an intervention can be extremely dependent on the time scale of interest (Fig. 1.2). For example, investments in adaptation are the only real way of reducing impacts for the next 30–40 years, because the benefits of mitigation are realized with a lag of roughly this length (see Chapter 2). However, nearly all assessments agree that adaptations are less effective than mitigation for reducing impacts by 2100.

Much of this book will focus on time scales of the next few decades, rather than the end of the century. There are several pragmatic reasons for this choice. First, growth in food demand is expected to be faster before 2050 than after, because global population growth will likely decelerate. According to the United Nations' medium growth rate scenario, for instance, population will increase by 2.8 billion people between 2000 and 2050 but only by 0.2 billion between 2050 and 2100 (Table 1.1). The challenges to food security of rising global demand are therefore

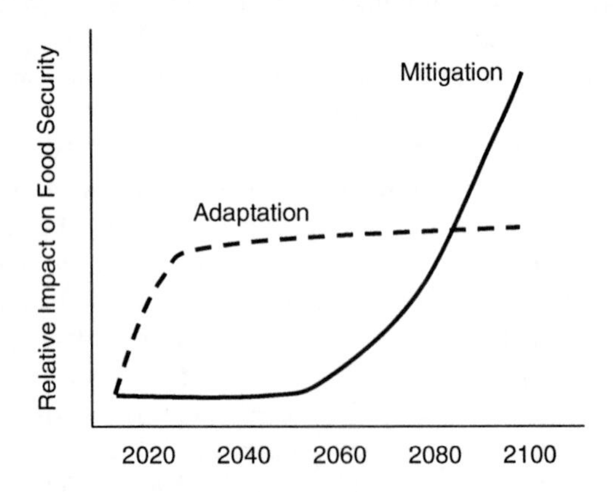

Fig. 1.2 A schematic view of the effectiveness of adaptation and mitigation as a function of time

Table 1.1 Projections of global population (billions of people) to 2050 and 2100 for different growth scenarios (data source: United Nations Population Division 2004)

Year	Low variant	Medium variant	High variant
2000	6.1	6.1	6.1
2050	7.4	8.9	10.6
2100	5.5	9.1	14.0

likely to be greatest in the near-term. Put differently, the challenge of the next few decades is to foster a tremendous growth in agricultural production in the face of climate change in order to improve food security, while after 2050 lies the more modest challenge of maintaining existing production levels in a warming world.

A second and no less important reason is that uncertainties in projections beyond 2050 are far greater than those of the next few decades. That is, it makes sense to focus on the aspects of the problem where projections are most likely to be accurate. Beyond 2050, agricultural technologies may be completely different from the current ones, and temperatures at a given location will often be beyond anything currently experienced (Chapter 2), making projections of climate impacts on agriculture very difficult.

Third, our experience is that most decisions, whether made in public or private sectors, do not account for time scales beyond a few decades. Therefore, scientific assessments of the near-term are likely to have a greater impact on societal choices than are those focused on the end of the twenty-first century, even if the latter would play a greater role in an ideal world.

Finally, and perhaps most importantly, is the fact that food security impacts of climate change in the next few decades may be severe in some locations. There is tremendous need for science that supports effective adaptations, especially when considering that most investments in agriculture take over a decade to provide substantial returns. Efforts to adapt to climate in 2020 or 2030 must therefore begin soon. There are obviously larger potential impacts as one looks beyond 2050, but one need look no further than the next 20 years to find a major scientific and societal challenge.

Although we believe these reasons are compelling enough to justify a focus on the short-term, we recognize that a view of the longer term is also needed, if only for the critical job of assessing the eventual benefits of mitigation. We therefore also include many references to assessments of 2080–2100, although none to our knowledge extend beyond this time.

1.4 Overview of Book

This book is organized to cover the major topics that, in our opinion, are needed to address this interdisciplinary subject. Part 1 of the book includes essential background information on food security and trends in the climate system. Chapter 2 presents an overview of food security and the potential ways that it can be influenced by climate changes. Chapter 3 presents an overview of climate models and how they are used to assess uncertainties in the future of climate.

Part 2 delves into the links between climate and crop yields, with a focus on how and how well scientists attempt to quantify these links with models and experiments. Chapter 4 describes the process-based ecophysiological crop models that underlie many assessments of yield impacts, while Chapters 5 and 6 discuss statistical approaches that model yields based on relationships gleaned from historical yield variations across time and space. Chapter 7 focuses on crop responses to elevated CO_2, a critical factor for projecting future yield impacts.

Part 3 provides an in-depth look at different adaptation options that may play a key role in minimizing impacts over the next few decades. Chapter 8 reviews market-mediated adaptations through trade, as well as on-farm agronomic adaptations such as shifts in planting dates or crop varieties. Chapter 9 discusses the key role that crop development can play in generating new seeds that can better thrive in a new climate.

Finally, Part 4 takes a snapshot of current research, with Chapter 10 summarizing results from recent regional and global assessments and Chapter 11 focusing on major lessons from past work and suggestions for future research.

1.5 Missing Pieces

Editors must inevitably choose to draw the line somewhere in order to balance the scope and focus of a book. The decisions made here reflect partly the expertise and interests of the authors involved, and partly our own biases on the most relevant and scientifically mature topics. However, we wish to point out many less mature topics that may prove important in the final analysis. For instructors who plan to use this book as a text, we encourage instructors planning to use this book as a text to supplement it with current papers on these topics.

Water Resources for Irrigation Impact assessments for irrigated regions often assume that water supply will be unaffected by climate change. Although this is a reasonable starting point (in order to focus first on the more direct effects of temperature and precipitation on crops), indirect effects through changes in regional water resources may be important. Studies that do link regional hydrology models to crop models, in order to simultaneously treat both supply and demand for irrigation water, have shown that local impacts and adaptation responses can be constrained by water supplies (Thomson et al. 2005).

Irrigation is currently practiced on roughly 17% of global cropland, with these systems contributing 40% of global food production (FAO 2002). Most of global irrigation water is applied in Asia, and therefore it is in this region that consideration of changes in water resources is most urgently needed. For example, it is widely acknowledged that much of the irrigation water in India and Pakistan originates as meltwater from Himalayan glaciers, that these glaciers are rapidly melting, and that summer streamflow may be significantly reduced within a few decades (Singh and Bengtsson 2004; Barnett et al. 2005; Rees and Collins 2006). Yet the implications of these limited water supplies on agriculture in general, and on the ability to adapt to climate change in particular, have to our knowledge only been superficially addressed.

Sea Level Rise Little work has considered the direct impacts of rising sea levels on agricultural production. Increases over the next few decades will likely be too small to have a major impact on agricultural production, but increases of more than 1 m, which are possible by the end of the century (Rahmstorf 2007), could result in the inundation of large tracts of low-lying coastal agriculture throughout Asia. Even smaller rises in the near-term could have strong local effects related to saltwater intrusion, with three particularly vulnerable sectors suggested in a recent FAO report: vegetable production, which tends to be irrigated in coastal regions, low lying aquaculture, and coastal fisheries (Bruinsma 2003).

Pest and Pathogens Farmers are constantly faced with the prospects of yield losses from weeds, animal pests, fungal and bacterial pathogens, and viruses. By one estimate, roughly 30–40% of global production for the major food crops is lost to these factors each year (Oerke 2005). Although climate change will undoubtedly modify pest dynamics, current understanding of these changes is quite limited (Easterling et al. 2007). With a few exceptions (Aggarwal et al. 2006), crop models in common use today do not include treatment of weeds, pests, or pathogens. Approaches to modeling responses to climate change include models that explicitly simulate weed competition or predator–prey interaction as well as simpler projections that use thresholds to define pest ranges. Nearly always the effects of temperature and CO_2 changes have been considered separately, although interactions between the two may prove important (Fuhrer 2003). Pests and pathogens can not only impact yields, but also the nutritional quality and health impacts of many crops. For example, carcinogenic aflatoxins are commonly found in maize and groundnuts and are most prevalent in hot and dry conditions (Chauhan et al. 2008).

Livestock and Fisheries This book focuses mainly on food crops, but meat, poultry, dairy, and fish are important sources of calories, protein, and income for many, including the food insecure. Livestock is a particularly important means of risk management (i.e. mixed crop–livestock systems) and adaptation to drought throughout much of the tropics (Thornton et al. 2007). Livestock systems broadly fit into two classes: those fed on grains or managed pasture grasses, such as in intensive feedlot systems common in developed countries, and those based mainly on grazing of wild grasses such as those common in poor countries with large malnourished populations. For the former, the main effects of climate change may be via crop yield and price changes discussed in this book, though higher temperatures will also present a challenge to management of heat stress and disease among animal populations. In pasture and rangeland systems, direct effects of heat on animals will be complemented by effects on forage quantity and quality. Pasture grasses in many temperate locations show yield increases for moderate warming, but also exhibit significant declines in nutrient content with higher CO_2 (Easterling et al. 2007).

In fisheries, interannual climatic variations, most notably related to the El Niño Southern Oscillation, lead to wide fluctuations in fish stocks. However, the net effects of future climate changes on fisheries are currently very uncertain, aside from a likely northward shift of many fish populations (Brander 2007). In addition to effects of warming, aquatic food webs could be as or more impacted by increased acidity resulting from oceanic CO_2 uptake (Easterling et al. 2007).

Mitigation in Agriculture Though this book focuses on the impacts of climate change on agriculture and food security, the role of agriculture in mitigating climate change is an important related topic. It has long been recognized that agriculture is a significant contributor to global greenhouse gas emissions, in terms of CO_2 and especially methane and nitrous oxide (Rosenzweig and Hillel 1998). Major reductions in emissions of these gases from agricultural activities could thus contribute to climate mitigation, and a myriad of technologies offer promise in this respect. For more information, a good starting point is the periodic reports of the Intergovernmental Panel on Climate Change.

Mitigation could even present an opportunity to adapt to climate impacts. For example, the prospect of a global emissions trading market will make it possible to generate rural income from either reducing emissions or providing renewable fuels. Such income, for example through biofuel production in poor oil-importing nations, may be an important means of income generation and represent a possible adaptation to declining staple crop production. Other synergies between adaptation and mitigation have been argued in the literature, such as the potential of conservation tillage practices to both sequester carbon in the soil and improve soil moisture needed in dry years (Lal 2004).

Policy Responses Despite the many uncertainties in physical and biological aspects of food security response to climate change, much of the inability to project future impacts relates to the simple fact that we cannot predict how humans will respond. Put differently, the severity of future impacts will depend in large measure on whether humans can effectively adapt. This book deals extensively with models of how rational farmers and regional economies might respond to climate change, but it should be clear that, like most other assessments, we implicitly assume that government policies that influence these behaviors remain fairly stable.

Any significant shifts in policy could dramatically affect the capacity of economies to cope with climate change, either for better or worse. One particularly important set of policies relates to long-term investments in the types of institutions and technologies that are needed to adapt to climate change, such as agricultural research or extension activities and emergency relief organizations. Funding for these activities has fluctuated in recent years and it is difficult to predict the future trajectory of overall policy support for agricultural development. Although many have argued convincingly that these investments offer high returns even in current climate (Alston et al. 2000), it can be difficult to prioritize long-term investments in public goods, particularly in poor countries.

Standing as a complement to these decisions about longer-term institutional investments are policies that deal with the short-term supply shocks that occur in years of bad harvests, shocks that may become more frequent and widespread with climate change. The recent experience with rapid price changes in 2008 provides a clear example. Many governments instituted new policies aimed at stabilizing local markets, including price controls, import tariffs, and export restrictions (FAO 2008). Yet the effect of these inward looking policies was often to destabilize global markets even further, causing rapid spikes in food prices and declines in food access in many rice importing countries.

How will governments respond if a year with extreme heat waves reduces global cereal harvests by 10% in 2020? Will they preserve existing policies and increase support of famine relief organizations, or will they embark on politically popular but potentially harmful protectionist policies? Though the recent experience of 2008 provides a cautionary note, perhaps it was a good learning experience that will lead to improved coordination during a future crisis.

There are many difficulties in predicting the future course of human decisions, not the least of which is that human behavior is not necessarily rational. As Bertrand Russell wrote: "It has been said that man is a rational animal. All my life I have been searching for evidence which could support this." Progress in anticipating future policy responses will therefore likely be slow.

The only certainty is perhaps that good policies, or the absence of bad policies, will be critical to maintaining food security in a changing climate. Identifying what these particular policies should be, and how to implement them, is beyond the scope of our book, but a topic that surely deserves much study in years ahead.

References

Aggarwal PK, Banerjee B, Daryaei MG, Bhatia A, Bala A, Rani S, Chander S, Pathak H, Kalra N (2006) InfoCrop: a dynamic simulation model for the assessment of crop yields, losses due to pests, and environmental impact of agro-ecosystems in tropical environments. II. Performance of the model. Agric Sys 89(1):47–67

Alston JM, Chan-Kang C, Marra MC, Pardey PG, Wyatt TJ (2000) A meta analysis of rates of return to agricultural R&D: Ex pede Herculem? International Food Policy Research Institute, Washington, DC, p 148

Barnett TP, Adam JC, Lettenmaier DP (2005) Potential impacts of a warming climate on water availability in snow-dominated regions. Nature 438(7066):303–309

Brander KM (2007) Global fish production and climate change. Proc Natl Acad Sci 104(50):19709–19714

Bruinsma J (ed) (2003) World agriculture: towards 2015/2030: an FAO perspective. Earthscan, Rome, Italy

Chauhan YS, Wright GC, Rachaputi NC (2008) Modelling climatic risks of aflatoxin contamination in maize. Aust J Exp Agric 48(3):358

Easterling W, Aggarwal P, Batima P, Brander K, Erda L, Howden M, Kirilenko A, Morton J, Soussana JF, Schmidhuber J, Tubiello F (2007) Chapter 5: food, fibre, and forest products. In: Climate change 2007: impacts, adaptation and vulnerability contribution of working group II to the fourth assessment report of the intergovernmental panel on climate change. Cambridge University Press, Cambridge, UK and NY, USA

FAO (2002) Crops and drops: making the best use of water for agriculture. FAO, Rome, p 28

FAO (2008) Policy measures taken by government to reduce the impact of soaring prices. http://www.fao.org/giews/english/policy/index.htm, April 1, 2009

Fuhrer J (2003) Agroecosystern responses to combinations of elevated CO_2, ozone, and global climate change. Agric Ecosyst Environ 97(1–3):1–20

Lal R (2004) Soil carbon sequestration impacts on global climate change and food security. Science 304(5677):1623–1627

Oerke EC (2005) Crop losses to pests. J Agric Sci . doi:10.1017/S0021859605005708: 1–13

Rahmstorf S (2007) A semi-empirical approach to projecting future sea-level rise. Science 315(5810):368–370

Rees HG, Collins DN (2006) Regional differences in response of flow in glacier-fed Himalayan rivers to climatic warming. Hydrol Process 20(10):2157–2169

Rosenzweig C, Hillel D (1998) Climate change and the global harvest. Oxford University Press, New York

Singh P, Bengtsson L (2004) Hydrological sensitivity of a large Himalayan basin to climate change. Hydrol Process 18(13):2363–2385

Thomson AM, Rosenberg NJ, Izaurralde RC, Brown RA (2005) Climate change impacts for the conterminous USA: an integrated assessment: Part 5. Irrigated agriculture and national grain crop production. Clim Change 69(1):89–105

Thornton P, Herrero M, Freeman A, Mwai O, Rege E, Jones P, McDermott J (2007) Vulnerability, climate change and livestock–research opportunities and challenges for poverty alleviation. SAT eJournal 4(1):1–23

United Nations Population Division (2004) World population in 2300. United Nations, New York

Chapter 2
Climate Effects on Food Security: An Overview

Marshall Burke and David Lobell

Abstract There are roughly 1 billion food insecure people in the world today, each having this status because food is unavailable to them, because it is unaffordable, or because they are too unhealthy to make use of it – or some combination of the three. Assessing the potential effects of climate change on food security requires understanding the underlying determinants of these three aspects of food security – availability, access, and utilization – and how climate change might affect each. This chapter explores these aspects and determinants of food security, summarizing the basic mechanisms by which climate change might impact the lives of the global food insecure.

2.1 Introduction

Roughly a billion people around the world live their lives in chronic hunger, and humanity's inability to offer them sustained livelihood improvements has been one of its most obdurate shortcomings. Although rapid improvements in agricultural productivity and economic growth over the second half of the twentieth century brought food security to broad swaths of the developing world, other regions did not share in that success and remain no better off today – and in some cases worse off – than they were decades ago.

Progress in understanding *why* some of these countries emerged from poverty and food insecurity, and why others did not, has been similarly limited. Such questions are central to the economics discipline and have been an active area of research for centuries, but they have generated remarkably little consensus on how to effect the transition from poverty to wealth.

Much of the controversy arises because food security (and related measures of well-being) have multiple, complex determinants, with varying agreement on which causes are more or less important. But confronting this complexity is central to any understanding of the potential impacts of climate change on food security. For

M. Burke (✉) and D. Lobell
Stanford University, CA, USA

D. Lobell and M. Burke (eds.), *Climate Change and Food Security*,
Advances in Global Change Research 37, DOI 10.1007/978-90-481-2953-9_2,

instance, knowledge of the impacts of climate on crop yields alone is not enough to understand food security impacts, because food security is a product of complex natural and social systems in which yields play only one (albeit important) part. Instead, understanding climate change's full impact will require knowledge of its potential effects on both the proximate causes of food insecurity (e.g., low agricultural yields, low rural incomes) as well as on the more fundamental causes of poor economic progress (e.g., poorly-functioning institutions and markets, low education levels, high disease burden). Our goal in this chapter is not to assign priority among possible factors, but to outline how each might be affected by climate change and what in turn this could mean for progress towards achieving global food security.

2.2 Food Security: Definition, Measurement, and Recent Progress

Although an earlier study counted at least 30 definitions of the term "food security" (Maxwell and Smith 1992), the benchmark understanding of the term is roughly that of FAO (FAO 2001):

> Food security is a situation that exists when all people at all times have physical, social, and economic access to sufficient, safe and nutritious food that meets their dietary needs and food preferences for an active and healthy life.

Under this definition, food security consists of having, on an individual level, the food one needs and wants. This definition is then conventionally subdivided into three main components: *food availability, food access*, and *food utilization. Availability* refers to the physical presence of food; *access* refers to having the means to acquire food through production or purchase; and *utilization* refers to the appropriate nutritional content of the food and the ability of the body to use it effectively. We explore each of these aspects of food security in the context of climate change below.

2.2.1 *Measuring Food Security*

Proper measurement of food security is of clear policy and humanitarian concern, primarily because such measures are used to both assess progress in a given region and to target assistance where needed. However, given the multiple interacting components of food security listed above, measurement of food security is both difficult and controversial.

The most cited country- and global-level statistics on food security are those of FAO, who use a measure of "undernourishment" as a proxy for food security. This measure relies primarily on national level data on food supply to estimate the percentage of a given country's population that does not have access to sufficient dietary energy. FAO's estimation procedure, shown graphically in Fig. 2.1, is roughly as follows (Naiken 2002):

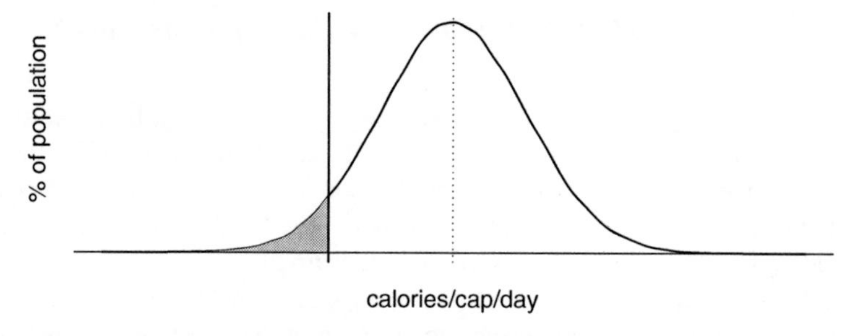

Fig. 2.1 Schematic of FAO undernourishment calculation. *Dotted line* = average per capita calorie availability; *curve* = population distribution around that average; and *vertical black line* = calorie undernourishment threshold. *Grey shading* represents proportion of population that is undernourished (after Naiken 2002)

1. For a given country, sum up the total number of calories available for human consumption in a given year, which will be a combination of locally produced food and imported food, minus exports.
2. Divide by the country's population to determine average per capita consumption.
3. Determine the shape of the distribution around this mean, either from household income or expenditure surveys where available, or from imputation from other sources where not.
4. Use country-level data on average height and weight to estimate the minimum amount of energy needed to maintain light activity, and apply the distribution from (3) to determine what percentage of the population falls below this threshold.

This undernourishment measure is attractive because it is both computationally simple and based on relatively available national-level data on the production and trade of agricultural products. But many criticize the measure for effectively focusing on food availability at the expense of issues of household food access and utilization – the status of which might correlate poorly with the national-level estimates of food supply on which the FAO measure is based. Others complain that the FAO statistics reveal nothing about the sub-national location or severity of food insecurity, and thus that they are of little use to practical policy planning (Smith et al. 2006).

Recent work by Smith et al. (2006) seeks to address these concerns by using nationally representative household survey data to construct detailed measures of food security. This approach tallies up the amount of food each household reports purchasing or producing, and based on these totals calculates how many households fall below given calorie and diet quality thresholds. Although more difficult and time-consuming to construct, and more limited in their spatial and temporal coverage given the absence of household surveys for many developing countries and many years, such measures can provide much more detail on both the nature and location of food insecurity in a given country. Furthermore, as discussed below, this approach often yields different conclusions about the severity of food insecurity than the benchmark FAO measures.

2.2.2 Where and How Numerous Are the Food Insecure?

Progress in reducing the number of food insecure over the last half century is at once both promising and discomforting. As Fig. 2.2 shows, since 1970 there has been a general decline in both the number of global food insecure and their percentage of the total population, as calculated using the FAO undernourishment measure described above. These reductions were driven primarily by large gains in East and Southeast Asia, where decades of strong economic growth liberated hundreds of millions from poverty and food insecurity. In both of these regions, the prevalence of undernourishment fell from 40% to 45% in 1970 to near 10% in 2004.

These remarkable gains stand in contrast to two more worrying trends. First, progress in reducing global food insecurity seems to have slowed and even reversed in the last few years, with the number of global food insecure actually rising slightly for the last two years for which there are data. Second, Sub-Saharan Africa stands out as a region for which progress has been particularly discouraging. While South Asia continues to have the highest total number of food insecure (around 300 million by the undernourishment measure), SSA is gaining rapidly and has the highest prevalence of food insecurity at around 35% of the population – a rate that has shown little deviation over the last 4 decades.

Moreover, household survey-based estimates of food insecurity suggest that FAO statistics might underestimate the prevalence of food insecurity in the region. Using estimates of food insecurity based directly on household survey data for 12 African countries, Smith et al. (2006) calculate rates of food insecurity on average 20% higher than FAO estimates, rising up to as much as 40% in some countries (Table 2.1), and attribute much of the difference to significantly lower estimates of mean food consumption when using household survey data directly. Household data also suggest differences in the relative rates of hunger across the same sample

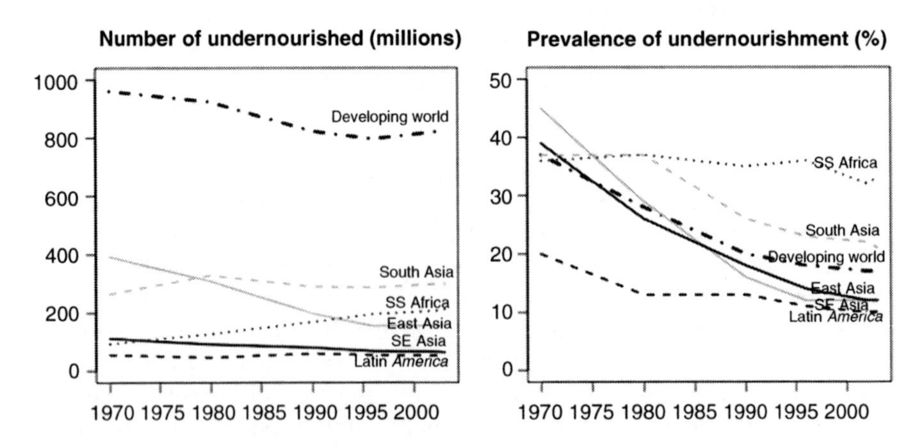

Fig. 2.2 Regional trends in undernourishment, 1970–2004. *Left panel*: number of undernourished in millions. *Right panel*: percent of undernourished in population (FAO 2009)

Table 2.1 Comparison of FAO and household survey-derived estimates of the prevalence of food energy deficiency for 12 African countries (from Smith et al. 2006)

Country	FAO estimate	Household survey estimate	FAO rank	Household survey rank
Ethiopia	44	76	4	1
Burundi	66	75	1	2
Malawi	32	73	8	3
Zambia	45	71	3	4
Rwanda	41	65	7	5
Mozambique	63	60	2	6
Senegal	24	60	10	7
Ghana	15	51	12	8
Guinea	31	45	9	9
Kenya	43	44	5	10
Tanzania	43	44	6	11
Uganda	21	37	11	12
Mean	*39*	*59*		

Table 2.2 Prevalence of rural and urban food energy deficiency in selected African countries (from Smith et al. 2006)

Country	Rural prevalence	Urban prevalence	Percent of food energy deficient who are rural	Rural population as % of total population
Burundi	76	41	95	90
Ethiopia	74	90	82	85
Ghana	50	53	53	55
Guinea	40	54	59	66
Kenya	46	30	71	62
Malawi	73	76	84	84
Mozambique	63	51	70	65
Rwanda	67	55	86	83
Senegal	54	69	45	51
Tanzania	42	53	60	66
Uganda	36	41	86	88
Zambia	71	71	65	65

countries, which is potentially of relevance to policy-makers trying to target assistance priorities across countries (right columns of Table 2.1). For instance, Ethiopia ranks as the fourth most food insecure country in the sample using FAO data, but the most food insecure country using household data.

Household data also allow further insight into the location of poverty within countries. While gripping images of urban slums are often the public face of food insecurity, household data typically reveal that the majority of the food insecure reside in rural areas. Table 2.2 shows that while the prevalence of food insecurity can be as high or higher in urban areas, a much greater percent of the total number of food insecure in a given country live in rural areas, largely reflecting the much higher percentage of the total population still residing in rural regions. And although the developing world is urbanizing, broader analyses of survey data suggest that the majority of the world's poor and food insecure will remain in rural areas for years to come (Ravallion et al. 2007).

This basic picture of the state of global food security – strong recent progress in some regions, little progress in other regions, many of which remain desperately poor, and the dominant role of rural populations in the total number of food insecure – provide the baseline for our exploration of the effects of climate change on the three aspects of food security, which we now take up in turn.

2.3 Food Availability and Climate Change

The *food availability* dimension of food security encompasses issues of global and regional food supply, and asks the basic question: can we physically produce enough food to feed our population? There is a vast literature on past trends and future trajectories in the world's ability to feed itself which cannot be adequately summarized in the current chapter (Conway and Serageldin 1997; Dyson 1999), Nevertheless, any discussion of the effect of climate change on the global food supply must take into account current realities and trends in global and regional supplies of food. We therefore highlight three particularly important characteristics of the global food supply.

The first is that on an average per capita basis, the world today produces more than enough food to meet caloric requirements, and that this success has been based mostly on yield gains over the last half century. Perhaps first popularized by Thomas Malthus in the early 1800s, the question of whether the world can produce enough food to feed a growing population has been a perennial concern. Thus far, technology has mostly precluded Malthusian doomsday predictions of population-driven food shortages. Through the first half of the last century, the need for increased food production was met by expansion of cropped area. But beginning in about the 1950s, when population and income growth were adding increasing pressure to global food markets, large-scale sustained investment in crop productivity greatly increased yields of crops throughout the developing world. This so-called Green Revolution allows the world today to produce 170% more cereals on just 8% more cropped area than 50 years ago (Panel (a), Fig. 2.3) – certainly an incredible achievement. Furthermore, on a global level this productivity growth has more than kept pace with the large observed increases in population, and global per capita cereal production currently stands at almost exactly 1 kg/person/day – or more than enough, on average, to feed everyone on the planet.

These global averages, however, hide large regional discrepancies, and the second important characteristic of the global food supply is that there are stark regional differences in the magnitude and source of agricultural productivity growth – differences that provide important insights into the challenge a changing climate might pose. Panels (a–c) in Fig. 2.3 show area, yield, and per-capita production trends by region

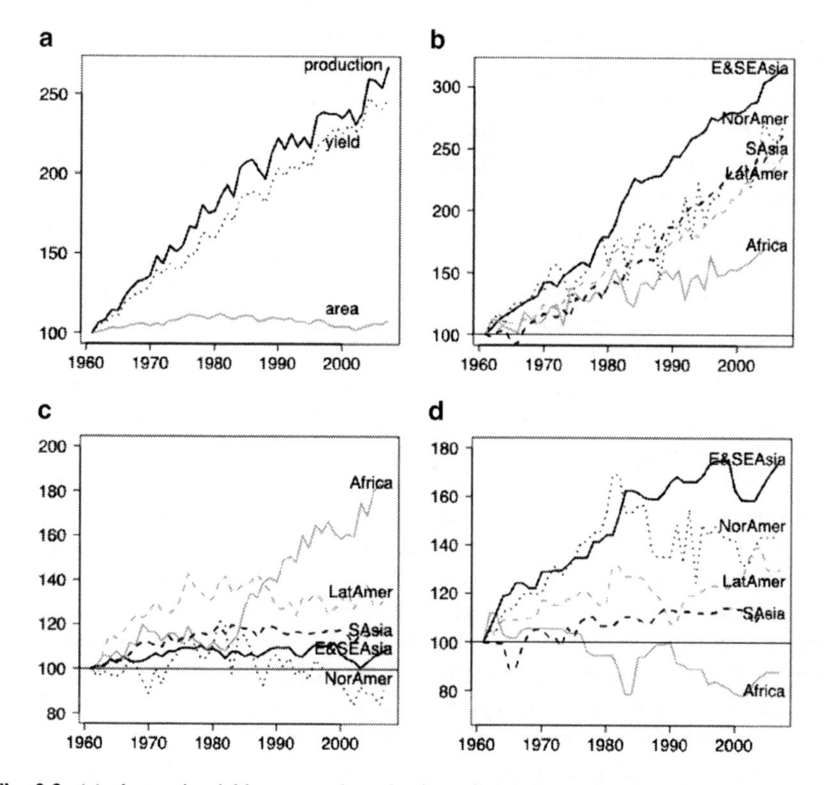

Fig. 2.3 (**a**) change in yield, area, and production of global cereals, 1961–2007. (**b**) Regional yield trends and (**c**) area trends over the same period. (**d**) Changes in per capita production. All values are indexed (1961 = 100) (FAO 2009)

over the last half-century. While most regions in the developed and developing world enjoyed somewhere between a doubling and tripling of yield since 1960, allowing them to increase their per-capita production of cereals with only minimal expansion in cropped area, Africa stands out as the continent on which progress has been most difficult. African cereal yields have grown at less the half the Asian rate, and despite an 80% increase in the amount of cropped area on the continent, total cereal production has not kept pace with population growth. As a result, the African continent is the only region where per capita production of cereals has declined over the last half century.

The potential for reversing this decline and for further boosting productivity elsewhere in world is at once promising and troubling. The promise for Africa and other low productivity regions lies in the large gulf between observed yields and potential yields – the so-called "yield gap" – much of which is explained by low adoption of modern agricultural technology and inputs. In theory, developing appropriate agricultural technology for these regions and providing the proper incentives to use it could rapidly close these substantial yield gaps and quickly raise productivity. But elsewhere in the world, particularly in the high-input systems in much of North America, Europe, and parts of Asia, yield gaps are much smaller, and achieving the sustained increases in yield observed over the past 4 decades will likely be very difficult without further increases in yield potential ceilings (Cassman 1999).

Furthermore, expanding cropped area, which is the alternative to increasing yield, is either difficult or unappealing throughout much of the world, either because of urban encroachment on agricultural land or because of the environmental costs of bringing new land into production. The FAO, which periodically assesses trends in crop demand and supply, envisages a significant expansion of cropland area in Africa and Latin America but little growth elsewhere, mainly because so little land in Asia remains uncultivated (Bruinsma 2003). Overall, most global assessments project that (1) crop demand will grow considerably over the next few decades, given the additive pressures of population growth (estimated to peak at 9.1 billion mid-century), higher incomes resulting in shifting food preferences, and potential development of large-scale biofuel production and the additional crop demand it represents; (2) the rate of demand growth, however, will be slower than observed in the past few decades, as population begins to stabilize; (3) and based on existing land, water, and fertilizer resources, crop production should be able to keep pace with the decelerating demand growth, but only with a formidable and sustained investment in yield improving technologies, cropland expansion, and input use.

The third important feature of the global supply situation is that food is now a truly global commodity, and the movement of food across borders plays an increasingly important role in meeting regional food demand. As Fig. 2.4 shows, about 10% of world cereal production is traded internationally, with some regions (Oceania, North America) exporting substantial amounts of what they produce, and other regions (notably Africa) importing up to a third of what they consume. Such food trade can either buffer or exacerbate the effects of a local food supply shock. A country experiencing drought, for instance, might make up for production shortfalls through imports, but cereal

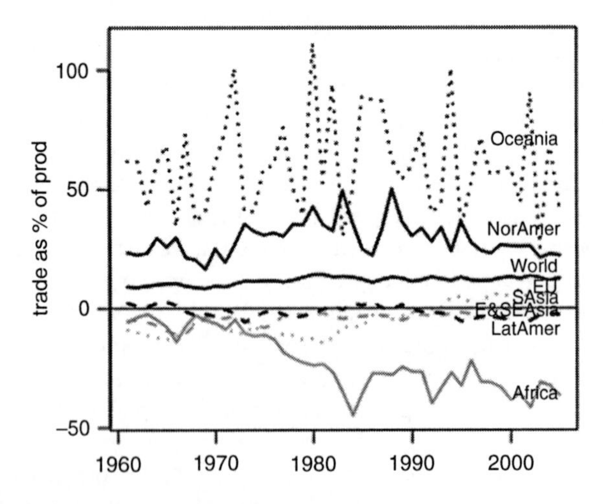

Fig. 2.4 Trade in cereals as a percent of production, 1961–2005. Regional values are exports or imports as a percent of production, with negative values indicating net importers. Global values reflect total trade as a percentage of production (FAO 2009)

importing nations would pay higher food prices on the world market when large exporting countries suffer similar shortfalls. In the event that such shocks happen simultaneously, poor importing nations would need to import when prices are very high, greatly increasing their difficulties in bolstering local food supplies.

So how might climate change affect global and regional food supply? As the rest of this book will show, climate change will have potentially large effects on both agricultural yields and potential cropped area, with global trade acting as a potential buffer when countries trade and when climate shocks are not uniform across space. But agricultural production and food availability are just one part of the food security story, and we now turn to the less frequently discussed potential effects of climate change on access and utilization.

2.4 Food Access and Climate Change

If Thomas Malthus is the customary jumping-off point for discussions of food availability, economist Amartya Sen dominates introductory paragraphs in discussions of food access. Recalling the definition above, food access refers to the ability of an individual to acquire food, either through its production or its purchase. Sen referred to these means of food acquisition as "entitlements", and he won the Nobel Prize in part for showing how famines were a result of households or entire regions periodically lacking entitlements. His basic insights hold today: for a farmer, entitlements are the means of food production available to her (e.g., land and labor), and her access to food is secure if she can command sufficient amounts of these factors to produce enough food. For those who don't farm, access to food is a function of incomes and prices – how much money one has to spend on food, and how much the food costs. Food access then can deteriorate when non-farm incomes fall, when food prices rise, or when the productivity of farm households suffers.

Determining the effects of climate change on food access for a given household therefore requires addressing the role of climate change in relation to four basic questions: how households earn their income, the nature of their exposure to food prices, how well integrated their local food markets are with global markets, and their broader longer-run prospects for livelihood improvement.

The first question concerns *the extent to which a given household is dependent on agriculture for its income*. If agriculture will be one of the sectors most affected by climate change, then the greater a household's livelihood depends on agriculture, the more that household is sensitive to the impacts of climate. While good systematic data on sources of household income in the developing world are hard to come by, there have been multiple recent efforts to try to systematize the available survey data on household income and to discern basic patterns across the developing world.[1]

[1] See, for example, the RIGA project (Davis et al. 2007, Ivanic and Martin 2008), IFPRI's HarvesChoice Project, Stanford's ALP Project, and Banerjee and Duflo (2007).

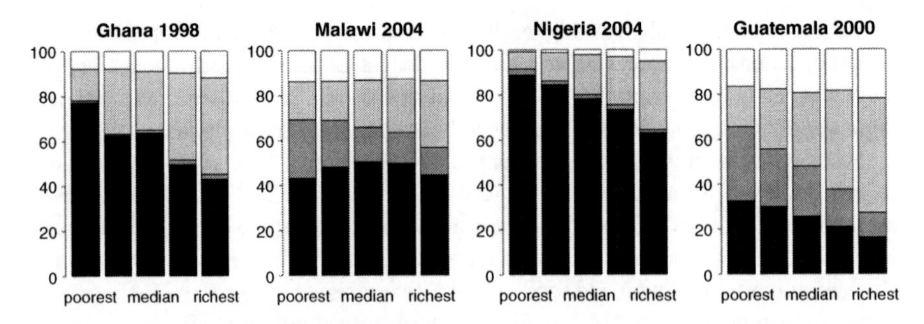

Fig. 2.5 Percentage of rural household income derived from agricultural sources, by income quintile for selected countries (Davis et al. 2007). *Black* = on-farm income; *dark grey* = agricultural wage income; *light grey* = off-farm income; *white* = transfer income/other. For instance, the poorest 20% of rural Ghanian households derive about 80% of their income from farm activities

Recall from the Smith data that most of the food insecure live in rural areas. Figure 2.5, adapted from data in Davis et al. (2007), shows the percentage of rural household income derived from agriculture in a set of poor countries for which good income data were available. The general trend from these data is clear: rural households in many developing countries depend to a significant extent on agriculture for their livelihoods, and this dependence tends to rise the poorer the household is. For the poorest of these households, two-thirds or more of income is earned on average through agriculture – a total that includes income from sales of crop and livestock goods in the marketplace, as well as the value of such goods produced by the household for home consumption. Such an agricultural dependence suggests that the income effects of a decline in agricultural productivity (all else equal) could be significant.

Importantly, however, few households even in rural areas are fully dependent on agriculture. The inherent seasonality and year-to-year variability of agriculture encourages diversification of income sources, and in the dry season or in particularly bad years many rural households seek additional income in non-agricultural wage labor or self-employment. As Fig. 2.5 shows, these sources of income can be important, and introduce a second main aspect of climate change and food access, *the nature of a household's exposure to food prices.*

All households are consumers of food, and as consumers benefit when food prices are low. But rural households are often producers of food as well, selling surplus in local markets. As a result, such households benefit as consumers but are hurt as producers when food prices fall. So if climate change induces changes in the supply of food that in turn affect food prices, the net impact of these price changes on food access in a given household will depend on the particular net consumption position in that household – that is, whether they spend more on food purchases than they earn from selling what they produce.

Estimating net consumption position again requires the use of household surveys, in this case surveys that have detailed information on both agricultural production and consumption behavior. As with income, there have been some recent efforts to characterize household net position for staple grains across a subset of developing countries (Fig. 2.6). These data show, unsurprisingly, that urban households are largely net consumers of food, purchasing nearly all of what they consume. More surprising perhaps is that the majority of rural households in many poor countries are also net consumers of food, with even farm households using non-farm income to purchase what they are unable to produce. These net-consuming households will likely be helped if prices fall, or hurt if climate change makes food more expensive.[2]

Finally, the extent to which these net consuming households are affected by changes in food prices depends on how much of their income they spend on food, and on what types of food they buy. For instance, most households in wealthy countries are substantial net consumers of food, but because they spend such a small percentage of their total income on food, they are little affected if the price of food changes. This is not the case in poorer households, who can spend half or more of their income on food (Fig. 2.7), and for whom changes in food prices can have serious effects on the quantity and quality of food consumed. Because climate change might also affect the relative prices of different staples (for instance if warming hurts one cereal more than another), the particular diet composition of

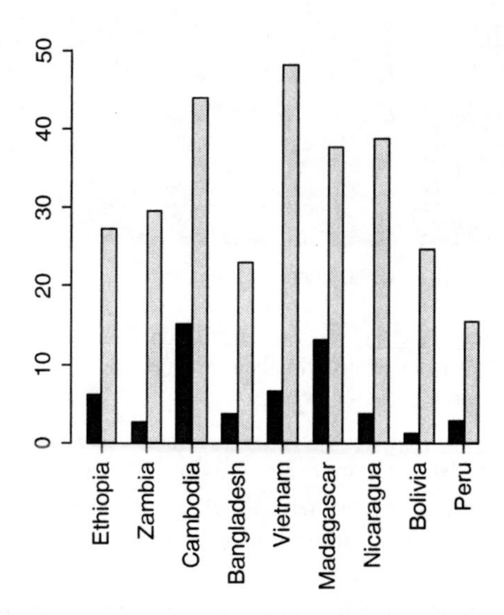

Fig. 2.6 Percent of households who are net sellers of staple crops, selected countries (Ivanic and Martin 2008). *Dark grey* = urban households; *light grey* = rural households

---

[2] There are cases where the longer-run effects of high prices might actually benefit net consumers, for instance if in response to the incentives of higher prices they are able to expand their own production and become net sellers of food, or if higher food prices induce expansion of production on other farms and raise the total demand for agricultural wage labor. For a more complete treatment of these longer-run dynamics, please see Singh et al. (1986).

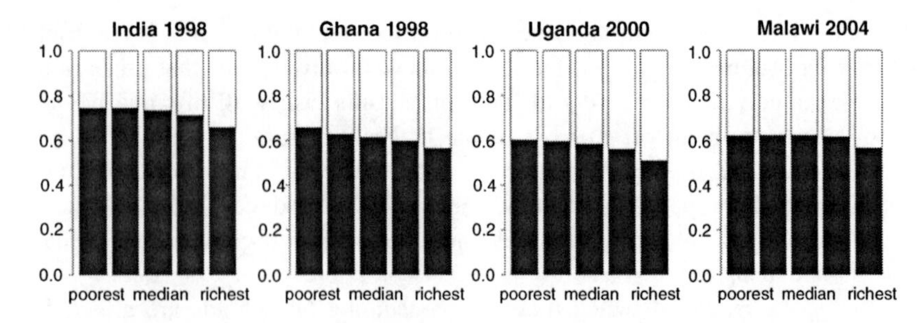

Fig. 2.7 Food expenditure as a percent of total household expenditure, by expenditure quintile. India data are for Bihar and Uttar Pradesh. Data from Stanford's ALP project (Karen Wang, pers. comm.)

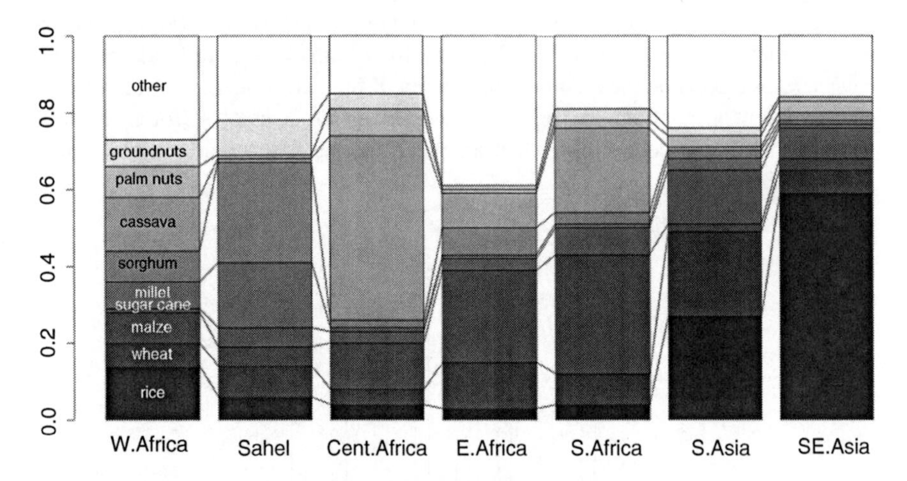

Fig. 2.8 Average percent of dietary calories derived from different crops for selected regions. Data are from FAO (2009), as calculated in Lobell et al. (2008)

poor households can also be important. As Fig. 2.8 shows, this composition can vary greatly from region to region, with the three primary cereals (rice, wheat, and corn) accounting for over 75% of calories consumed in parts of Asia, to less than 20% throughout much of Africa.

The third important determinant of climate change's effects on food access concerns *how well integrated local food markets are with global markets*. As discussed in later chapters, the effects of climate change on agricultural productivity will likely vary by region, and so it is important whether in a given area local food prices and availability are driven primarily by local shifts in production, or whether that area is well integrated with regional or global food markets such that local prices track global price movements. This degree of integration could play a large role in the welfare effects of climate in a given region. For instance, a region that suffers large productivity losses under climate change but whose food markets are well integrated with global markets could see little change in the price of food if it is able to import food to cover losses. Conversely, a country well integrated with global markets could see food prices rise even if it doesn't experience local climate effects.

The final determinant of the effects of climate change on food access concerns *the degree to which longer-run prospects for growth in income and food security are climate sensitive*. This question is undoubtedly the most contentious of the four, because there is remarkably little agreement on the underlying causes of economic development, and thus little understanding of the relative importance of climate in determining why some countries become rich and others remain poor over the long run.

The economics literature offers perhaps three main explanations of why some countries have succeeded economically over time and others have not (Easterly and Levine 2003). The first explanation, argued prominently by Bloom and Sachs (1998) among others, suggests that *geography* is central to long-run economic success. Noting the high correlation between tropical location and underdevelopment, proponents of this explanation argue that a country's geographic location directly shapes various factors fundamental to long-run economic success – for instance the quality of the country's soils, the favorability of its climate for agriculture and habitation, the prevalence of various diseases, and the ease with which goods can be traded within and across its borders.

A second strain of thought places primary emphasis on the role of *institutions* in economic development. This explanation, promoted by Acemoglu et al. (2001) and Easterly and Levine (2003) among others, argues that economic progress has less to do with a country's soils and climate and much more to do with the quality of its institutions – in particular, factors such as limited corruption and institutional respect for private property and the rule of law.

A final explanation focuses on the role of particular *policies* in explaining long-run economic performance. Proponents in this camp (Williamson 1990) argue that even with favorable geography and well-functioning institutions, countries with bad economic policy are destined for poor economic growth. They point to instances in which poor economic management resulted in the collapse of otherwise prosperous countries as evidence of the primacy of good policy.

A casual observer might suspect that all three explanations – geography, institutions, and policies – play some role in shaping long run economic success. But if one explanation is relatively more important than another – a possibility that each camp adamantly claims is the case – then climate change could have a greater or lesser effect on longer run prospects for the alleviation of poverty and hunger. In particular, if the climate worsens, and it is in fact geography that constrains eventual economic success, the aggregate effects of climate change on food security could be great. If on the other hand institutional quality dominates long-run success, then climate change could have little effect on long-run progress.

Aside from these important questions about the long run determinants of economic progress, however, it should be clear that climate plays an important and direct role in the immediate food security of a large number of the world's poor. For households who eat much of what they produce, or who face food prices tightly linked to local agricultural production – and these households number in the hundreds of millions – the welfare effects of a negative supply shock can be large and lasting. Various studies demonstrate the persistent welfare effects of short-term adverse climate shocks for rural households, as for instance households in crisis sell productive assets to meet immediate consumption needs (Dercon and World Institute for Development Economics 2002; Hoddinott 2006). If climate change alters the likelihood of these shocks, we could expect large effects on rural household welfare in poor countries, even if the economy-wide consequences are minimal.

2.5 Food Utilization and Climate Change

Even if climate change were to have minimal impacts on the supply of food or on the ability of households to access it, it could still affect food security through its effects on the *utilization of food*. The utilization component of food security is perhaps its murkiest and least well-studied aspect, but generally relates to the nutritional aspects of food consumption. Supposing availability and access issues are taken care of, achieving proper food utilization requires satisfactory answers to three questions: does the food an individual eats contain all the energy, protein, and nutrients necessary for her to lead a healthy and productive life? Is the food itself safe and not likely to make her ill? And finally, is the individual healthy enough to take advantage of the food's nutritional qualities?

New evidence is indeed emerging about the potential effects of climate change on food utilization. Nevertheless, and as in the case of food access, climate will be only one component of a broader suite of issues that shapes an individual's ability to utilize food properly.

2.5.1 Food Utilization and Nutrition

Although a primary purpose of food is provision of dietary energy, and widely used undernourishment indicators such as FAO's lean heavily on estimates of calorie consumption to estimate food security trends, food is of course much more than just energy. Food also provides protein and various nutrients essential for bodily function, and there is increasing recognition of the important role insufficient intake of these nutrients plays in global illness and death from infectious disease (Black 2003). Importantly, prevalence of micronutrient deficiencies around the world is generally higher than estimates of caloric deficiencies, and alleviating these deficiencies has become a major public health priority.

Table 2.3 lists major micronutrient deficiencies, some of their health effects, and the most recent estimates of their global prevalence. It reveals that estimated prevalences for deficiencies in nutrients such as iodine and zinc are more than twice the FAO benchmark estimates for number of global undernourished. As a result, added together these micronutrient deficiencies account for one of the largest sources of global health loss (Lopez et al. 2006).

Table 2.3 Global prevalence of micronutrient deficiency (http://www.who.int/vmnis; (Ezzati 2004))

Micronutrient	Effects of deficiency	Number of global deficient (billion)	Percent of population deficient (%)
Iron	Child and maternal mortality, reduced cognitive development	1.6	25
Iodine	Reduced cognitive development, deformation, goiter	1.9	31
Vitamin A	Blindness, immune deficiency	0.6 (children <5 yrs)	20
		0.1 (women 15–44)	6
Zinc	Immune deficiency	1.9	31

Most poor households receive what micronutrients they do get through the consumption of plants, with vitamins sourced largely from fruits and leafy greens, and minerals from cereals. For instance, some estimates suggest that 80% of African and Southeast Asian intake of vitamin A comes through fruit and vegetable consumption (Ruel 2001). Meat and dairy products are a primary source of many nutrients in the developed world, but are often too expensive for poor households, and are thus a minor source of micronutrients throughout much of the developing world.

Climate change could directly affect micronutrient consumption in three main ways: by changing the yields of important crop sources of micronutrients, by altering the nutritional content of a specific crop, or by influencing decisions to grow crops of different nutritional value.

There is little published evidence on the effects of climate change on micronutrient content of crops, and also much less evidence on the potential effects of climate change on fruits and vegetable yields compared to that available for cereals. Some studies show that higher CO_2 concentrations can lower protein content in various food crops, particularly in the context of low nitrogen inputs (Taub et al. 2008). While the estimated reductions could be relatively modest in magnitude – 10–15% decrease in grain protein content by around the end of century – such declines would be amplified by any yield losses, and would hit hardest in poor areas where nitrogen application rates are low and where crops constitute a primary source of dietary protein.

Beyond direct effects on yields, climate can also shape the decisions farmers make about what crops to grow (Rosenzweig and Binswanger 1993), and thus could potentially alter planting decisions in ways that alter micronutrient availability. For instance, in the poor soils and highly variable climates of much of central and western Africa, starchy tubers such as cassava and yam often dominate cropping systems, in no small part because of their ability to achieve at least some yield in the worst weather years. Unfortunately, such crops are also very poor sources of both protein and micronutrients, and to the extent that they are favored in future climate relative to cereals as a source of dietary energy, nutrient consumptions could decline.

2.5.2 Disease and Food Utilization

Food utilization also concerns the ability of individuals to make use of the nutrients available to them, and is thus closely linked to both the overall safety of the food and to the individual's health. While not all unhealthy people are necessarily food insecure, health status can be a primary contributing factor to food security. Of particular concern in poor countries are the strong feedbacks between malnutrition and disease, in which undernutrition leads to increased infection and a higher disease burden, which in turn leads to energy loss, reduced productivity, and further diminished access to food (Schaible and Kaufmann 2007). And while the underlying determinants of health and food safety are complex and clearly extend far beyond narrow climate issues, most possible manifestations of climate change (e.g., warming, drought, or floods) have the potential to negatively affect health in ways that compromise food utilization (Confalonieri et al. 2007).

Growing evidence indicates the significant role climate can play in the safety of food, as pathogens enjoy warmer climates. For instance, warming temperatures have been shown to increase the incidence of *Salmonella*-related food poisoning in Europe and Australia, and warming ocean temperatures have been shown to increase the incidence of human shellfish and reeffish poisoning (Kovats et al. 2004; McMichael et al. 2006).

Perhaps more importantly, climate change has the potential to affect health status directly, in ways that alter an individual's ability to utilize food. In areas with limited access to clean water and sanitation infrastructure, diarrheal disease is a leading killer, and contributes directly to child mortality and poor food utilization by limiting absorption of nutrients. Extreme rainfall events, drought events, and warming temperatures have all been shown to increase the incidence of diarrheal disease, often significantly (Checkley et al. 2000; McMichael et al. 2006; Confalonieri et al. 2007). Warming temperatures will likely also expand the range of important vector-borne diseases such as malaria and dengue (e.g., McMichael et al. 2006). Similarly, changes in rainfall patterns could also affect disease incidence, for instance with increasing drought heightening the risk of meningitis outbreak, or increased extreme rainfall events increasing the likelihood of cholera outbreaks (McMichael et al. 2006; Confalonieri et al. 2007).

Unfortunately, all available evidence suggests that the health effects of climate change will hit hardest where disease burdens and susceptibility to disease are already high, and where public health infrastructure is poorly developed – that is, in the poorest countries of the world. And since diseases such as malaria and diarrheal disease disproportionately affect younger ages, the health burden of climate change will be borne primarily by children in the developing world. The broader food security impacts of these climate-related health losses have not been well quantified, and are a topic in immediate need of attention by researchers.

2.6 Summary

There are clearly many pathways through which climate change will impact food availability, access, and utilization. Climate induced changes in agricultural productivity will likely affect the incomes earned and the food prices faced by poor households, with the net effect on food security a function of each household's particular set of livelihood strategies. In addition, health impacts associated with climate change could hamper the ability of individuals to utilize food effectively. These multiple potential impacts will occur in the midst of broader trends in global and regional food security, which include rapid recent progress throughout much of the developing world, but little improvement across most of the African continent, much of which remains desperately poor and food insecure.

The remainder of the book treats in detail some of the evidence surrounding specific aspects of climate impacts on food security, and in particular the methods used to understand them. Somewhat inescapably, however, the book focuses on topics where current knowledge and methods are most developed – which are issues primarily surrounding the food availability aspects of food security. But this subsequent focus should not distract from the broader message of this chapter, which is that food security is more than just food production, and that some of the most important effects of climate on food security could be through its effects on incomes, food prices, and the health of the poor.

References

Acemoglu D, Johnson S et al (2001) The colonial origins of comparative development: an empirical investigation. Am Econ Rev 91:1369–1401

Banerjee AV, Duflo E (2007) The economic lives of the poor. J Econ Perspect 21(1):141–167

Black R (2003) Micronutrient deficiency: an underlying cause of morbidity and mortality. Bull World Health Organ 81(2):79

Bloom DE, Sachs JD (1998) Geography, demography, and economic growth in Africa. Brookings Pap Econ Act 2:207–273

Bruinsma J (2003) World agriculture: towards 2015/2030: an FAO perspective. Earthscan

Cassman KG (1999) Ecological intensification of cereal production systems: yield potential, soil quality, and precision agriculture. Natl Acad Sci 96:5952–5959

Checkley W, Epstein LD et al (2000) Effects of EI Ni–o and ambient temperature on hospital admissions for diarrhoeal diseases in Peruvian children. Lancet 355(9202):442–450

Confalonieri U, Menne B et al (2007) Human health. In: Parry ML, Canziani OF, Palutikof JP, van der Linden PJ, Hanson CE (eds) Climate change 2007: impacts, adaptation and vulnerability. Contribution of working group II to the fourth assessment report of the intergovernmental panel on climate change, Cambridge University Press, Cambridge, UK, pp 391–431

Conway G (1997) The doubly green revolution: food for all in the 21st century. Penguin, 334 pages

Davis B, Winters P et al (2007) Rural income generating activities: a cross country comparison. ESA Working Paper, Rome, FAO, p 68

Dercon S (2002) Income risk, coping strategies, and safety nets. World Bank Research Observer 17(2)141–166

Dyson T (1999) World food trends and prospects to 2025. Natl Acad Sci 96:5929–5936
Easterly W, Levine R (2003) Tropics, germs, and crops: how endowments influence economic development. J Monetary Econ 50(1):3–39
Ezzati M (2004) Comparative quantification of health risks: global and regional burden of disease attributable to selected major risk factors. World Health Organization, Geneva
FAO (2001) The state of food insecurity in the world. Rome, Food and Agricultural Organization of the United Nations, p 58
FAO (2009) FAOSTAT online database. http://faostat.fao.org. Retrieved 10 Jan 2009
Hoddinott J (2006) Shocks and their consequences across and within households in rural Zimbabwe. J Dev Stud 42(2):301–321
Ivanic M, Martin W (2008) Implications of higher global food prices for poverty in low-income countries. Policy Research Working Papers. New York, World Bank
Kovats RS, Edwards SJ et al (2004) The effect of temperature on food poisoning: a time-series analysis of salmonellosis in ten European countries. Epidemiol Infect 132(3):443–453
Lobell DB, Burke MB et al (2008) Prioritizing climate change adaptation needs for food security in 2030. Science 319(5863):607–610
Lopez AD, Mathers CD et al (2006) Global and regional burden of disease and risk factors, 2001: systematic analysis of population health data. Lancet 367(9524):1747–1757
Maxwell S, Smith M (1992) Household food security: a conceptual review. In: Maxwell S, Frankenberger T (eds) Household food security: concepts, indicators, measurements. IFAD and UNICEF, Rome and New York
McMichael AJ, Woodruff RE et al (2006) Climate change and human health: present and future risks. Lancet 367(9513):859–869
Naiken L (2002) FAO methodology for estimating the prevalence of undernourishment. Methods for the measurement of food deprivation and undernutrition. FAO, Rome
Ravallion M, Chen S et al (2007) New evidence on the urbanization of global poverty. Popul Dev Rev 33(4):667–701
Rosenzweig MR, Binswanger HP (1993) Wealth, weather risk and the composition and profitability of agricultural investments. Econ J 103:56–78
Ruel MT (2001) Can food-based strategies help reduce vitamin a and iron deficiencies? A review of recent evidence. International Food Policy Research Institute (IFPRI), Washington, DC
Schaible UE, Kaufmann SH (2007) Malnutrition and infection: complex mechanisms and global impacts. PLoS Med 4(5):e115
Singh I, Squire L et al (1986) Agricultural household models: extensions, applications, and policy. Johns Hopkins University Press, Baltimore, Maryland
Smith LC, Alderman H et al (2006) Food insecurity in sub-Saharan Africa: new estimates from household expenditure surveys. International Food Policy Research Institute, Washington
Taub DR, Miller B et al (2008) Effects of elevated CO_2 on the protein concentration of food crops: a meta-analysis. Glob Change Biol 14(3):565–575
Williamson J (1990) What Washington means by policy reform. In: Williamson J (ed) Latin american adjustment: how much has happened. Institute for International Economics, Washington, DC

Chapter 3
Climate Models and Their Projections of Future Changes

Claudia Tebaldi and Reto Knutti

Abstract This chapter describes global climate models and their output. The current approaches for analyzing their simulations, characterizing the range of likely future outcomes, and making projections relevant for impact analysis are described, specifically referring to the latest assessment report of the Intergovernmental Panel on Climate Change. We provide a summary of future projections of average temperature and precipitation changes at continental scales, together with a broad brush picture of the likely changes in indices of extremes, characterizing both temperature and precipitation events. An analysis of changes in growing season length is also presented as an example of climate model output analysis directly relevant to studies of climate change impacts on food security.

3.1 Where Do Climate Change Projections Come from?

Humans are conducting an unprecedented, deliberate yet uncontrolled experiment using our planet as its subject. Human-induced emissions of greenhouse gases and other pollutants, together with changes in land use, like deforestation, are altering our climate system properties in ways that are already detectable (Hegerl et al. 2007). The experiment is continuing, with future emissions projected to steadily raise the greenhouse gas concentrations in the atmosphere. This is because greenhouse gases like CO_2, unlike other gas species have a long life measured in decades and centuries, so that emissions over the years accumulate and increasingly alter the natural state of the system.

C. Tebaldi (✉)
Climate Central, Princeton, NJ, USA
email: ctebaldi@climatecentral.org

R. Knutti
Institute for Atmospheric and Climate Science, ETH (Swiss Federal Institute of Technology), Zurich, Zwitzerland

D. Lobell and M. Burke (eds.), *Climate Change and Food Security*,
Advances in Global Change Research 37, DOI 10.1007/978-90-481-2953-9_3,

Because we are changing the natural climate state like never before, it would be unreliable to simply extrapolate current trends into the future in order to predict what we will experience as a result. This is particularly true when we focus on regional changes, which are most important for devising adaptation measures. The interactions and reactions of the system are too complicated to be approximated by statistical models. In fact, as we will see, they are often complicated enough to present a challenge even for process-based, dynamical climate models. Rather, climate scientists use numerical models to construct surrogates of the real system, in order to perform a controlled, and replicable, version of the experiment. In this fashion they can test different assumptions in future anthropogenic emissions and other parameters regulating the climate system, span a wide range of uncertainty at least with regard to the known unknowns, and thus offer a range of climate change scenarios attempting to span a substantial portion of the relevant uncertainties.

There exists a hierarchy of climate models, from simple energy-balance models that can only approximate the trajectory of global mean temperature to models of intermediate complexity (Claussen et al. 2002) that can only resolve very large regions, to global coupled models, which are the subject of this chapter. These extremely complex computer models, also called atmosphere–ocean general circulation models (GCMs), divide the surface of the Earth, the depths of the oceans and the layers of the atmosphere into grid boxes. These GCMs describe the evolution of a host of climate variables at each grid box and for various time steps (between a few minutes and an hour) by solving differential equations derived from well-established physical laws, such as conservation of energy and angular momentum.

In the typical climate change experiment the simulation starts from conditions representative of the climate of pre-industrial times (around 1850), and is performed by letting the system evolve according to the laws of physics, undisturbed (i.e. not prescribing any observed changes), except for so-called external forcings to the system. Some of these external forcings occur naturally, like changes in solar irradiance (the 11-year solar cycle for example) or volcanic eruptions, which may be energetic enough to spew large quantities of aerosols in the stratosphere. The volcanic dust acts as a reflective cloud, partially shielding the surface of the Earth from incoming radiation and thus having a short-lived cooling effect on the order of a few years.

Particularly important in climate change experiments are increasing atmospheric concentrations of greenhouse gases, which are another form of external forcing, but anthropogenic rather than natural. They are imposed according to standard scenarios agreed upon by the scientific community, reflecting hypotheses about the future evolution of socio-economic, technological and political factors. Climate model simulated changes are therefore termed projections rather than predictions, because they are usually conditional on the assumed storyline or scenario. The system responds to these protracted anthropogenic forcings by altering its behavior in a trend-like fashion, rather than by cyclical or episodic changes which are typically the result of natural disturbances. These changes can be assessed by analyzing the output of a GCM experiment which is typically at least two and a half centuries long, producing simulations of climate from pre-industrial conditions out to the end of the twenty-first century, and taking on the order of weeks to be carried out on super-computers at research centers around the world.

As both our scientific understanding of climate process and our computing power improve, more and more processes at increasingly finer scales can be represented explicitly in these simulations. The size of a GCM grid box is limited by the amount of computer power available. Doubling the resolution of a model grid, for example going from 250 km by 250 km grid boxes, typical of the current models, to 125 km by 125 km grid boxes makes the model about ten times slower to run. Even with relatively fine resolutions there always remains the need for approximating those processes that act at scales not explicitly represented. It is these approximations that are the source of large uncertainties, since many of the fine scale processes are responsible for the physical feedbacks that ultimately determine the direction and size of the changes of the system in response to its perturbations. Furthermore, fine scale processes are critical in determining the statistics of climate at local scales, which are usually the most relevant in determining impacts.

Let's consider a concrete example. The typical resolution of the GCMs that will participate in the next (fifth) assessment report of Intergovernmental Panel on Climate Change (IPCC, whose latest assessment report on the physical science basis of climate change is Solomon et al., 2007) will consist of about 200 km-wide boxes. An important process not explicitly represented at these scales is cloud formation. Nevertheless, the model needs to answer questions such as: how large a portion of the box is covered by clouds, given the temperature, humidity, pressure and wind conditions simulated at the box scale? What kind of clouds are going to form, high or low? How does the presence of aerosols influence the water holding capacity of cloud particles? How many water droplets will form, and what is the threshold for rain? The answers to these questions at each time step of the simulation are governed by parameters in the equations whose values are best guesses informed by experiments and observations, but contain a measure of uncertainty which reverberates in space and time within the simulation. Because the parameterizations are describing the large-scale effect of the cloud rather than actually resolving the processes in the clouds, the values used in the parameterizations often need to be chosen to match some observed evidence, but they do not represent real physical quantities that can be measured directly with any instrument. The effect of clouds on temperature and of course precipitation behavior, and the ensuing interactions among climate variables, is extremely significant and determines the magnitude of the changes simulated in response to external forcings.

Different GCMs are developed across the world. About 15 research groups of different nationalities have produced climate models which use different solutions to the numerical integrations, grids of different resolution, different sets of processes explicitly represented (does the model have interactive vegetation? Interactive carbon cycle?) and, most importantly, different approximations to the unresolved processes. What results is an ensemble of models which could be thought of as a set of best guesses, and can help address the question of structural uncertainty across models.

However, within a single model, formulations of many alternative parameterizations of sub-grid scale processes and parameter values are admissible, and costly experiments that vary those settings and thus explore within-model sources of uncertainty are being performed as well, albeit in just a handful of modeling centers because of the resources that they require. These are called perturbed-physics experiments, and probably the most famous example is given by climateprediction.net (Stainforth et al. 2005) whereby tens of thousands of variations of a Hadley Centre GCM (developed by the UK MetOffice) are distributed to personal computers all around the world, which run the model experiment in their idle time and send back results to a group of scientists in Oxford, who then analyze them to determine to which combination of changes in parameters' value the model is most sensitive.

Because different models make different choices about which processes to model and how to model them, there is a clear need to explore climate change projections across sets of GCMs, rather than relying on a single model's results. Also important are the limitations inherent in the resolutions of global models, which limit the models' abilities to represent local climates accurately, especially when those climates are influenced by complex topography not accurately represented at the GCM resolution. The limitation in resolution also undermines the models' ability to simulate particular sets of variables. Precipitation – especially summertime precipitation that is caused by small-scale convective processes – is a typical example. Winds at the surface are another example. As a result, there exists a cascade of confidence in the output of GCMs among climate scientists and modelers. Smooth fields of temperatures at continental scales are considered fairly reliable, details of temperature at regional scales less so. General tendencies in precipitation – changes given as a function of latitudes, for example – are generally agreed upon, but local features much less. In general, large area averages are considered more reliable than spatial details, and mean values are more robustly represented than variability and trends (e.g., Räisänen 2007).

Nevertheless, impact analysis needs regional detail. In order to "translate" large scale projections to local scales, two techniques of so-called "downscaling" are used. Regional dynamical models covering a limited domain can be nested into global models. Alternatively, statistical relations between the large scales and local scales may be derived on the basis of observations and applied to the large-scale projections. A simple and common example of this approach is to use only GCM projections of *changes* in temperature or precipitation, rather than absolute values, and add these changes to historical weather data from local stations.

In both dynamic and statistical downscaling, spatial detail is added to the coarse grid scale results, but part of that information is often just interpolation rather than providing additional knowledge and understanding. Regional models can be run down to ten or fifty kilometers, because they are run for a limited area (for example, North America, Europe, or South Africa) and usually for limited simulation times (for example 20 years at the end of the twentieth century, 20 years straddling 2050 and 20 years at the end of the twenty-first century). Similarly, statistical relationships can be fitted very economically between point-locations (like weather stations) and large scales and then applied to GCM output. Often, through the statistical approach, bias corrections or other calibration of the model output to the statistics of the observed regional weather, like variance inflation, can be imposed. Dynamical downscaling output has been shown to reproduce the statistics of extremes more accurately, thanks to the higher resolution at which simulations are conducted. However, the limitations and (most importantly) the uncertainty inherent in the results of the GCM used to drive the regional downscaling are inevitably passed down to the regional results. In order to address the characterization of uncertainty across models, similarly to what is being accomplished by coordinated experiments at the GCM level, efforts to conduct systematic downscaling from an ensemble of global models are being made, and some programs are well under way or are planned to be associated with the next IPCC report activities. (e.g., the North American Regional Climate Change Assessment Program; the PRUDENCE program in Europe, described in Christensen and Christensen 2007; Vrac et al. 2007).

There are other sources of uncertainty when it comes to future projections, mainly natural variability and emission uncertainty. Natural variability is due to the chaotic nature of weather processes which determine fast fluctuations in the time series of any given variable of interest. By definition climate is the long-term average behavior of the system, and in this respect fast fluctuations cancel out. However, when running impact models that need daily data, for example, it is important to feed different realization of model simulations based on different initial conditions to get a measure of the natural variability at play. For some variables (e.g., seasonal mean temperatures, averaged over decades) the uncertainty due to natural variability becomes secondary compared to the uncertainty due to modeling and emission scenarios. For other variables (e.g., precipitation, or extremes), natural variability maintains an important role in the overall uncertainty of future projections. Uncertainty in the magnitude of future greenhouse gas emissions is driven by uncertainty in the socio-economic, technological and political factors that will determine population growth, technological progress, energy demand and so on. Scientists have so far refused to assign probabilities to different scenarios of greenhouse gas emissions, opting for designing standard pathways of future emissions (Nakicenovic and Swart 2000), so that model experiments can adopt these prescribed alternative storylines and so their results can be analyzed conditionally on a specific emission scenario. We show in Fig. 3.1 time series of CO_2 concentration levels from three of the most commonly explored SRES scenarios: A2, A1B and B1.

One should be careful in mixing scenario and model uncertainty, as the two are quite different. In a sense, scenario uncertainty is a matter of choice when making decisions, whereas the modeling uncertainty reflects our limited understanding or incomplete description of the true climate system in a numerical code. In the remainder of this chapter we then focus on modeling uncertainties.

As a summary of the discussion so far we list in Box 3.1 the main sources of uncertainties, their causes and their possible solutions.

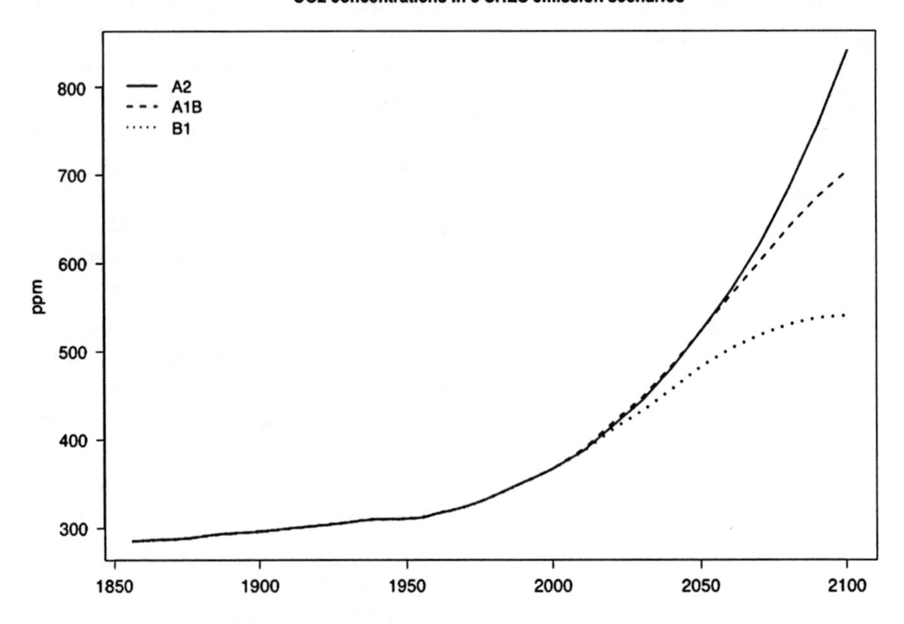

Fig. 3.1 CO_2 concentrations along the centuries as prescribed by three commonly used SRES scenarios: A2 (*solid line*), A1B (*dashed line*) and B1 (*dotted line*). Units on the *y*-axis are parts per million (ppm)

3.2 Utilizing GCM Projections

Given that there is no easy alternative to producing future projections based on GCMs, we now delve deeper into the way uncertainty manifests itself in GCM projections, and what the attempts are at reducing it or at least characterizing and quantifying it in robust ways. Figure 3.2 shows changes along the twentieth and twenty-first centuries in global temperature from the climate models (21 of them) used in IPCC AR4 that performed their simulations under the emissions scenario SRES A1B (the set of models that have contributed experiments to IPCC AR4 is also known as the World Climate Research Program Coupled Model Intercomparison Phase 3, a.k.a. CMIP3, Meehl et al. 2007a). Clearly, some models warm more rapidly than others. There is consensus across models that the future will be warmer but even at the global average scale the difference in the magnitude of warming is large, with up to a factor of 2 between the two extremes of the range. The same observation applies to the trajectories of global mean precipitation (not shown).

The uncertainty increases when we consider regional changes. Two maps in Figs. 3.3 and 3.4 show geographic patterns of change by the end of this century

Box 3.1: Sources of uncertainty in climate change projections

1) Initial conditions: slight changes in the starting point of the simulation change where the wiggles in the trajectory happen, due to the natural variability of the system (i.e., the chaotic nature of weather). In order to account for this, ensemble of runs by the same model/under the same scenario in which only the initial state of the system is varied are used to characterize the range of natural variability, and are averaged when looking at climate statistics, making the dependence from initial conditions disappear.
2) External forcings: different pathways of greenhouse gas and aerosol emissions cause very different evolutions (trajectories) of the perturbed climate system. As a consequence numerous different pathways need to be explored and adaptation policies tested against possible alternative futures. Notice though that for short-term projections the outcome is very similar no matter what the emission scenario is. Most of what will happen in the next two or three decades is the result of "commitment", based on what we have emitted so far.
3) Unresolved or poorly understood system behavior: certain climate processes are not perfectly understood (i.e. ice sheet collapse mechanisms are still beyond our scientific grasp) or are not perfectly modeled (cloud behavior is not resolved and thus directly simulated by GCM, local weather patterns are not reproduced because of the coarse topography represented in these models). Increasing the resolution of models (which goes hand in hand with increasing computing power) and ultimately the progress of our scientific understanding will ameliorate this problem. Meanwhile, perturbed physics and multi-model ensembles help span the range of possible answers, and quantify this kind of uncertainty.

computed as the ensemble mean of the same set of models run under the same A1B scenario. The stippling in the figures marks points in space where 90% or more of the models in the ensemble agree on the direction of change (quite a lenient condition). Again, models agree that our planet will get warmer everywhere, but the agreement on the sign of precipitation change is not as strong, except for the high latitudes of the northern hemisphere and some areas of the tropics, expected to become wetter, and limited areas of the subtropics, expected to become drier.

What that means for a specific region is that a histogram of the average precipitation change (for a given season, or annually) may straddle the zero line. In fact for many areas the ensemble mean change is very close to zero, hiding a range of possibilities that go all the way from significant increases to significant decreases in the average quantity of rain falling in the future.

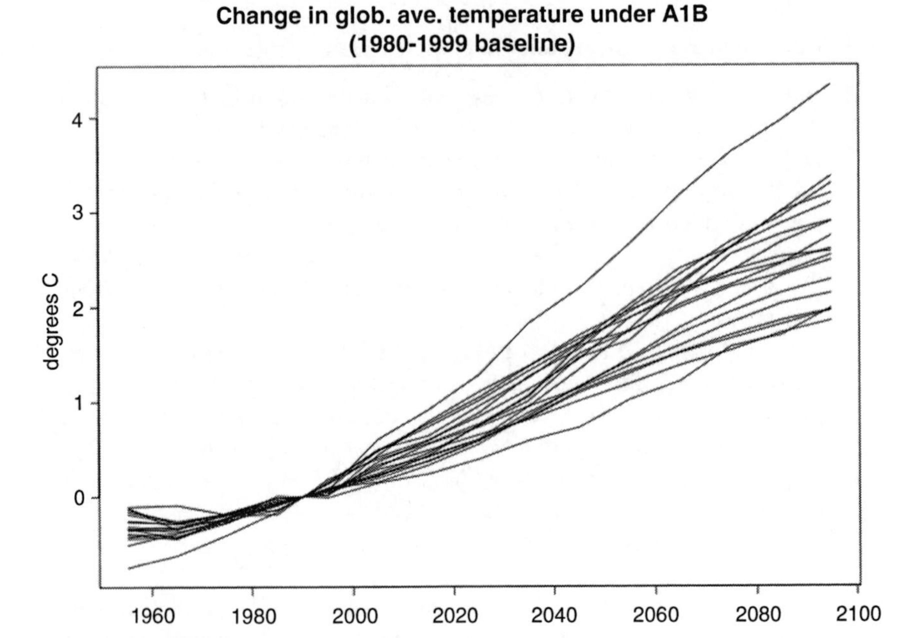

Fig. 3.2 Global average temperature changes along the twentieth and twenty-first centuries from a set of GCMs (18 models from the CMIP3 archive used in IPCC AR4) that performed their simulations under the emissions scenario SRES A1B and made available both temperature and precipitation output (for consistency with Figs. 3.2 and 3.3). Units are degrees Celsius, changes are with respect to the two decadal average 1981–2000. Each line corresponds to a different GCM. The trajectories are connecting 15 decadal averages

3.2.1 One Model, One Vote?

It should be obvious then that future scenario analysis based on a single GCM would be a dangerously narrow view of what could be possibly in store.

The IPCC report's future projections chapters, for global and regional projections (Meehl et al. 2007b; Christensen et al. 2007), have in fact adopted a multi-model approach. For the most part the results in the report consist of simple descriptive statistics of climate change across the ensemble. Maps of ensemble means, accompanied by measures of uncertainty like ensemble ranges or standard deviations, or simple measures of model consensus, like stippling indicating majority vote, are used to communicate projected changes especially when no specific region is in focus, and the aim is to paint a global picture of the future climate.

There is a justification for this one-model-one-vote approach. It can be found in the results of informal or formal assessments that have demonstrated how no model outperforms all others, when a comprehensive set of diagnostics are brought to bear (Gleckler et al. 2008). The same kind of analysis has demonstrated that the central

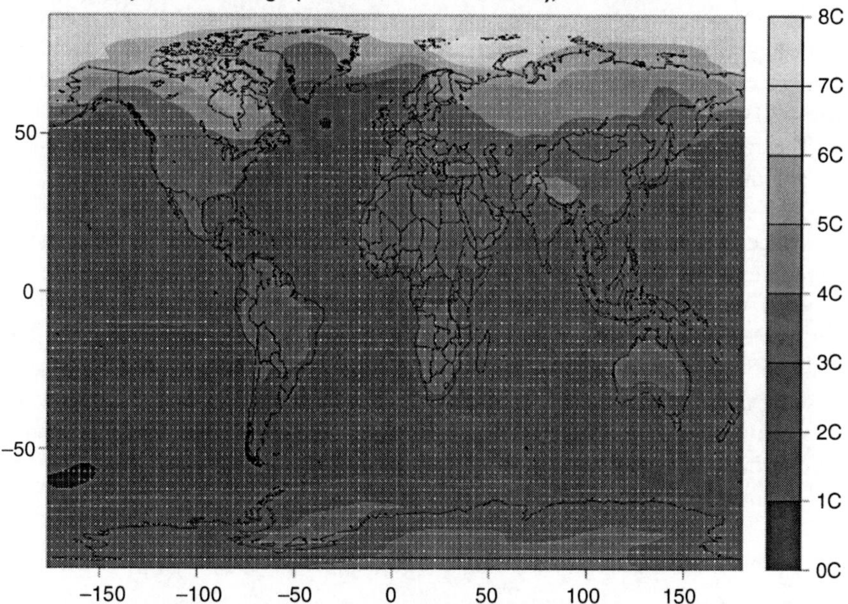

Fig. 3.3 Map of the geographic patterns of temperature change by the end of this century (2081–2100 vs 1981–2000) computed as the ensemble mean of the same set of models used for Fig. 3.1, and the same emission scenario experiment, A1B. Dots on the map mark grid boxes where 90% of the models agree over the sign of temperature change

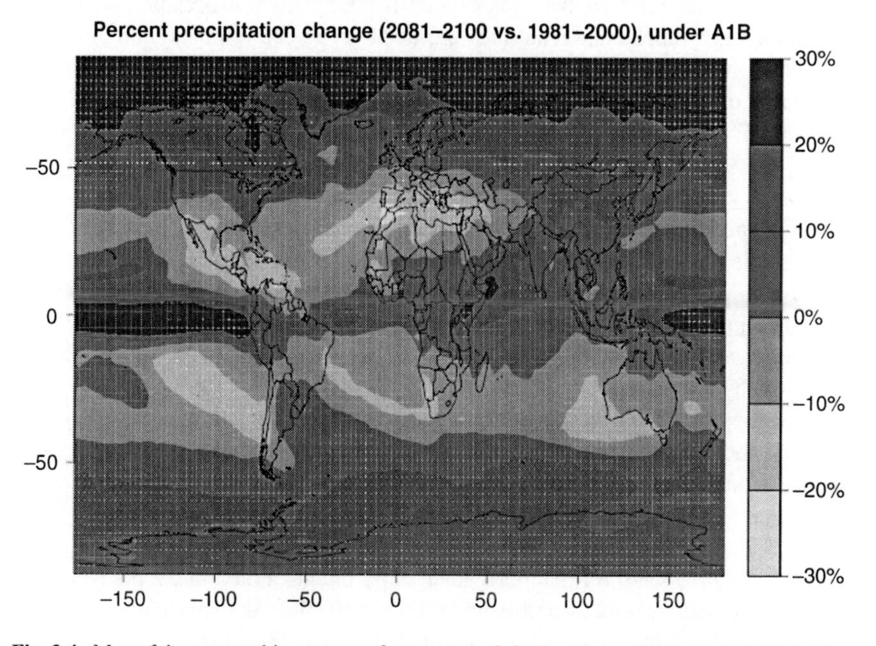

Fig. 3.4 Map of the geographic patterns of percent precipitation change by the end of this century (2081–2100 vs 1981–2000) computed as the ensemble mean of the same set of models used for Fig. 3.1, and the same emission scenario experiment, A1B. Dots on the map mark grid boxes where 90% of the models agree over the sign of precipitation change

tendency of the models, when evaluated over the current part of the integration and compared to observations, outperforms any single model simulation (Reichler and Kim 2008).

The crucial point is that if we take a multifaceted approach to validation, and if we are interested in mean climatological quantities, no model is the true model, and model means are a safer bet than any single model's output. Of course, good performance in reproducing current climate is a necessary condition for these model experiments but does not guarantee equally good performance for future climate simulations. In fact, simple tests that stratify models, and their future projections, with respect to the quality of their current simulations have shown that there is often no significant difference between the ranges of projected changes obtained by considering only "good models", or all models. The difficulty lies in defining a metric of what constitutes a good model in a situation where the same model is used to predict many different aspects of climate in different regions, and where no true independent evaluation of the forecast is possible.

3.2.2 Different Models, Different Votes?

From a pragmatic standpoint, in view of the increasing attention and activities in the area of adaptation, simple ensemble means and ranges have the desirable property of being easy to interpret so that non-experts handling multi-model projections can straightforwardly appreciate what they are dealt. The need remains though to alert users to some shortcomings of these multi-model ensembles. They have been called "ensemble of opportunity" for very important reasons: they are not intended to be a systematic exploration of uncertainties, there may exist dependencies among the models and systematic errors common to all of them, and there is no easy way to rank or pick and choose better and worse models. There is also a more general aspect of model projections that invites careful consideration. An important characteristic that sets climate model projections apart from other kinds of numerical forecasts (e.g., daily weather, or seasonal forecasts) is the lack of validation, since the projections usually consists of multi-decadal mean changes at some point far in the future, and are conditional to emissions scenarios that may not be realized exactly as hypothesized.

Nevertheless, GCMs remain our best guess at future changes, especially regional changes, and the existence of coordinated experiments by many modeling groups, willing to make their respective output available in public archives facilitates a cautious approach to model uncertainty, even if some sources of uncertainty remain elusive.

Since 2000, when the first coordinated experiments aimed at coupled-models' comparison made results available under the CMIP flag (the simulations that will be made available for the next IPCC report will be labeled CMIP5) formal statistical approaches to combining multi-model ensembles started to be developed and to appear in the literature (e.g., Räisänen and Palmer 2001; Giorgi and Mearns 2002, 2003; Tebaldi et al. 2004, 2005; Smith et al. 2009). Most of these approaches use a Bayesian paradigm, in order to provide probabilistic projections of quantities like temperature and precipitation change. Departing from a simple count of observed frequencies in the ensemble, these methods formally posit an initial best guess, i.e. prior distribution, of the quantities of interest (be those current and future climate variables, or the different models' weights) and use the data collected (observations and model simulations) to reshape it into so-called posterior distributions. This is done through Bayes theorem by writing down the likelihood of the data as a function of the unknown quantities and combining it with the prior distribution of the unknown quantities. The Bayesian paradigm offers a natural means of incorporating expert judgment, which is formalized in the prior probabilities (for example, scientists may be asked to specify ranges and the distribution of probability within them for unknown quantities, like climate sensitivity or model reliability). If no such information is available, prior distributions are chosen uniform over a large interval, or otherwise very diffuse, like Gaussians with very large variance.

For some of these methods the final result may not be significantly different from an empirical histogram of models' individual projections, but the formal nature of the derived probabilities may be considered of value if incorporated in quantitative risk assessment exercises, for example.

It is fair to say that these methodological developments are in their infancy, and each study accounts for some aspects of the peculiar nature of this problem, but each also makes some approximations. There is a method-dependent nature to their results, and different statistical approaches have been shown to deliver different estimates of the probabilities of interest (Christensen et al. 2007; Tebaldi and Knutti 2007). If one accepts the statistical assumptions of a given method, however, the propagation of the uncertainties to impact models is rigorously achieved. For example, in Tebaldi and Lobell (2008) a formal quantification of the uncertainty in temperature and precipitation projections at the regional scales through a Bayesian hierarchical model was used as input to a statistical model of crop yield changes for several staple crops, in order to derive probabilistic projections of changes in yields accounting for several sources of uncertainties (climate change, relation between climate change and crop change, CO_2 fertilization effect). We present the analysis in more detail in Section 3.3.2.

As discussed earlier, sometimes a simple descriptive analysis of ensemble model data is more interpretable. Sometimes it is the nature of the climatic variables of interest to pose obstacles to a formal statistical synthesis across models. Quantities like growing season length, or indices of climate extremes, are not as easily represented through statistical likelihood models as mean temperature or precipitation at large regional scales, for which a Normal distribution works in most cases. For these quantities and for the time being we may be better served by considering measures of model consensus and variability, like model spread, means and medians. We give an example of this kind of analysis in Section 3.3.1.

3.3 Two Examples: One-Model One-Vote and a Formal Bayesian Model

Here we explore two examples of analyses based on multi-model ensembles. We refer to the related papers for details on the actual results, aiming simply at juxtaposing two approaches that could be seen as spanning the methodology range of multi-model analysis.

We present first the approach from Lobell et al. (2008), which sought to rank 12 food-insecure regions in the developing world according to metrics of vulnerability in order to inform the prioritization of adaptation measures. The analysis uses projections of temperature and precipitation change by 2030 from 20 GCMs, part of CMIP3, and is performed by giving equal weight to each model simulation, computing summary statistics (medians and percentiles) of the ensemble simulations without applying any statistical synthesis of the climate projections first.

Our second example is the study by Tebaldi and Lobell (2008). Here the aim is to propose a formal probabilistic analysis of the impacts of climate change on the global yield of three important crops. In this case a statistical model combining the ensemble of simulations is used to derive joint probability distribution functions of temperature and precipitation changes, which are then sampled as input to the statistical crop model.

3.3.1 Descriptive Statistics of Multi-Model Data

In the paper by Lobell and co-authors (Lobell et al. 2008) an analysis aimed at prioritizing adaptation needs among 12 regions with currently marginal food security is based on climate projections from a CMIP3 ensemble of 20 models. Their simulations of temperature and precipitation change by 2030 under three alternative emission scenarios are extracted and averaged over each region. The empirical distribution of the models' signals of temperature and precipitation change is then sampled and the pairs of temperature and precipitation change are used as input of a statistical model of climate change impacts on crop yield for several basic crops, chosen because they are staples of the hungry's diet in each region.

The median projected impact of climate change on an important crop's production by 2030 and the 5th and 95th percentiles of the distribution of estimated impacts are used as indicators to form a measure of the vulnerability of the region (together with an assessment of the importance of the crop itself in the diet of the region's population). Since in the paper uncertainties in climate changes for the crop regions were quantified by randomly selecting joint changes in temperature and precipitation from the untouched population of models/scenarios projections, and feeding them through the estimated coefficients of the crop regression model (uncertainty in the crop regression model were also addressed by a bootstrap technique), this approach is an example of the one-model-one-vote approach, equally weighting projections from the

20 GCMs. The easy interpretability of the results for a multi-disciplinary audience is one desirable aspect of the analysis, trading off for more sophisticated approaches at combining projections into formal probability distribution functions.

3.3.2 A Formal Statistical Approach

The Tebaldi and Lobell (2008) paper is meant to be a methodological study. As the title suggests it proposes an approach towards a formal and rigorous quantification of the uncertainties that, from multiple sources, affect the estimates of climate change impacts in the agricultural sector.

A Bayesian hierarchical model is used to synthesize the joint projections of temperature and precipitation change from a multi-model ensemble, for a given SRES scenario. The output of this step of the analysis are bivariate probability distribution functions of future changes in temperature and precipitation, for the regions of the world where a given crop is cultivated, and tailored to an optimally defined crop-specific growing season. The next step of the method consists of sampling from these distributions pairs of change factors that are input to the statistical crop model (Lobell and Field 2007), similarly to what was done in Lobell et al. (2008) but substituting now a posterior distribution of climate changes to the empirical distribution of the CMIP3 models. The statistical treatment estimates the joint posterior probabilities by bringing to bear estimates of systematic biases in the models' simulations, estimates of the overall correlation of temperature and precipitation in the region and season analyzed, and observed trends in the two climate parameters and their degree of similarity to the simulated trends.

Like in any statistical modeling, assumptions on the data distributions are made and influence the final results, together with our assessment of the initial uncertainty in the quantities we want to estimate. This last point is a function of adopting the Bayesian paradigm, which updates a priori estimates of uncertainty through the information contained in the observed and simulated data. Nonetheless, the procedure provides a transparent and computationally efficient way of integrating the uncertainties at each step to make probabilistic statements about impacts, such as that by 2030 there is "larger than 80% chance that net losses for maize will exceed 10%" (Tebaldi and Lobell 2008).

3.4 Summary of Current Projections

Both temperature and precipitation output from all GCMs' twentieth century simulations have been found to be satisfactory representations of current climate in terms of mean geographical patterns, if analyzed at large scales (Räisänen 2007; Randall et al. 2007). Trend patterns are consistent with observations for those models that are forced by all known sources: volcanic eruptions, solar irradiance, greenhouse gases and aerosols (Barnett et al. 2005). Projected future warming patterns are robust (Meehl et al. 2007b), but global temperature change is uncertain by approximately 50% (Knutti et al. 2008) due to carbon cycle uncertainties (Friedlingstein et al. 2006) and models differing in their feedbacks and climate sensitivities (Bony et al. 2006, Knutti and Hegerl 2008). Short-term projections are better constrained by the observed warming than long-term projections (Knutti et al. 2002; Stott and Kettleborough 2002). This is because the effect of feedbacks amplifies with time and so do inter-model differences, so that differences across models become larger the farther in the future projections are. Models project changes in precipitation, extreme events (Tebaldi et al. 2006) and many other aspects of the climate system that are consistent with our understanding of climate processes and the consequences of a significant human influence on the climate, but agreement between models deteriorates as one moves from continental to regional to local (i.e. grid point) scales, and from mean quantities to more complex indices of climate events.

3.4.1 Temperature and Precipitation Projections by Region

Chapter 11 of the IPCC latest report by Working Group 1 describes in detail model projections for a set of subcontinental regions that have been traditionally used by the climate change community since they were proposed by Giorgi and Francisco (2000). The chapter also analyzes the processes relevant to each region's climate and the ability of models to capture them, thus gauging the reliability of future projections. It also considers the consistency of future projections with changes already observed, when possible, and supplements GCM projections by regional modeling studies when available. Figures, tables and discussion provide a rich portrait of what scientific understanding, local expertise and modeling experiments suggest for the future at these regional scales.

Here we describe continent by continent the main findings summarized by the IPCC report (Christensen et al. 2007). We intend this as a quick reference, but we point at the report chapter, available online with its own supplementary material, for a more complete treatment of the subject. The scenario adopted throughout the description is A1B, considered close to "business as usual", in the way its rates of emissions remain similar to the current rates. Temperature and precipitation projections are based on 21 models that ran the A1B experiment. Future changes are computed as the difference between two 20-year averages within each simulation, 1980–1999 and 2080–2099. In the case of precipitation the change is expressed as a percentage of the 1980–1999 average.

Interestingly, it has been shown (Santer et al. 1990) that regional patterns of change of temperature and precipitation remain close to constant along the future simulations, and the "intensity" of the change is proportional to the global average temperature change signal. This result, known as pattern scaling, has been exploited for example in order to explore a large range of uncertainties by modifying

Table 3.1 Global average temperature change (with respect to the baseline 1981–2000) under three SRES scenarios for short (2030) medium (2050) and long term (2070 and 2090) projections

SRES scenario	2021–2040	2041–2060	2061–2080	2081–2100
B1	0.9	1.1	1.5	1.8
A1B	0.9	1.5	2.2	2.7
A2	0.9	1.5	2.2	3.2

the parameters of simpler models, cheaper to run under alternative – but equally plausible – settings. From the simpler models only the signal of global average temperature change is extracted. It is then applied to the normalized geographic patterns derived from GCMs to produce a large collection of regional projections (Murphy et al. 2004). This argument may be used to infer shorter term projections on the basis of the following end-of-the century changes. To a first degree of approximation one simply computes the ratio of global average temperature change at the end of the century under the A1B scenario and the same quantity at the shorter projection time. The projected changes in temperature and precipitation can be then rescaled by dividing them by this ratio. As reference, the global average temperature change (with respect to the baseline 1981–2000) under the three scenarios for short (2030) medium (2050) and long term (2070 and 2090) projections are listed in Table 3.1.

3.4.1.1 Africa

The African continent will *very likely* (with greater than 90% probability, according to a rigorous definition of the phrase in the IPCC report) experience warming in greater measure than the global average, and this is true for all seasons. The median warming projected by the ensemble is of over 3°C, with individual model projections ranging from close to 2°C for the cooler models to over 5°C for the warmer models. The drier subtropical regions will warm more than the moister tropics. Annual rainfall changes will vary across the different regions of the continent. *Likely* (with greater than 2/3 probability) there will be a decrease of precipitation amounts in much of Mediterranean Africa and the northern Sahara, in southern Africa in the winter rainfall region and western margins. On the contrary it is *likely* that East Africa will experience an increase in annual mean rainfall. Projections for the Sahel, the Guinean Coast and the southern Sahara are of contrasting sign.

3.4.1.2 Mediterranean and Europe

Annual mean temperatures in Europe are *likely* to increase more than the global mean with the largest warming affecting Northern Europe in the winter season and the Mediterranean basin in the summer season. The median annual warming for Northern Europe is projected to be more than 3°C, with a range from more than 2°C up to almost 5.5°C. For southern Europe the median is higher, 3.5°C (range: 2–5°C).

Precipitation changes across models show a larger agreement than for other regions of the world, suggesting increases in Northern Europe especially in the winter season and decreases in southern Europe, largest in the spring and summer seasons.

3.4.1.3 Asia

The Asian continent will warm more than the global average almost in its entirety, the exception being South East Asia. The models project a median warming of over 3.3°C, with a gradient increasing towards the northern latitudes. The range of projections goes from over 2°C to well over 5.5°C, with seasonal ranges touching 8.7°C for winter in Northern Asia. Precipitation in the boreal winter season is projected to increase over the entire continent, with larger confidence in the Northern regions and the Tibetan Plateau. Precipitation in summer is also *likely* to increase in northern Asia, East Asia, South Asia and most of Southeast Asia, while models tend to agree over a decrease of precipitation in central Asia.

3.4.1.4 North America

The annual mean warming is *likely* to be greater than the global mean warming for almost all regions of the continent, but especially so for winter in the high latitudes (where minimum temperatures show largest increases) and summer in the Southwest (where maximum temperatures do). The median temperature change across the ensemble is above 4°C for the higher latitudes (Alaska and Canada) and above 3°C for the continental US region. The individual model projections range from close to 3°C as their minimum and over 7°C as their maximum for the northern portion of the continent, and from just above 2°C and up to 5.8°C for the lower tier. Annual mean precipitation is *very likely* to increase in Canada and the Northeast USA, and *likely* to decrease in the Southwest.

3.4.1.5 Central and South America

The annual mean warming is going to be *likely* close to the global mean warming in the southernmost part of South America (median warming of 2.5°C, range between 1.7°C and 3.9°C) but larger than the global mean warming in the rest of the region (median warming of above 3°C, range between 1.8°C and over 5°C). Annual precipitation is *likely* to decrease in most of Central America and in the southern Andes, but there is less confidence in the models being able to simulate the regional variability in these mountainous regions. Winter precipitation in Tierra del Fuego and summer precipitation in south-eastern South America is *likely* to increase. The agreement of models over annual and seasonal mean rainfall change over northern South America, including the Amazon forest, is poor, and does not allow to draw conclusions in a direction or its opposite.

3.4.1.6 Australia and New Zealand

Warming is *likely* to be comparable to the global mean, with the southern areas warming less, especially in winter. Median projection is 2.6°C in Southern Australia, 3°C in the Northern part (ranges between 2°C and 4.5°C). Decreases in precipitation are consistently projected for South and Southwest Australia, especially in winter and spring. Precipitation is *likely* to increase in the west of the South Island of New Zealand. Changes in rainfall in northern and central Australia are uncertain.

3.4.2 Extremes

Indices of climate extremes have been devised to extract information from GCM simulations beyond the behavior of mean quantities. We would not trust GCMs to simulate the statistics of extremes that we observe at local scales: quantities simulated by GCMs are intended as averages over the grid boxes that divide up atmosphere, oceans, and land of the GCM domain. Still, within each model's scale and climatology, indices of tail behavior can be analyzed for changes under increased greenhouse gas forcings. In this case too, ensembles of GCMs are used to draw conclusions regarding the consistency of changes across simulations, i.e. the degree of inter-model agreement. It is also the case that changes in the behavior of simulated extremes can be considered in light of observed changes, and scientific understanding. In the latter case, changes in processes that we are already observing, or should be expected in the future, are explained and understood in the context of a system perturbed by increasing concentrations of CO_2 in the atmosphere.

Many papers have recently addressed changes in extreme behavior. Here we briefly summarize some of our work that has specifically utilized GCM simulations. In Tebaldi et al. (2006) we analyzed five indices of extremes related to temperature:

- Frost Days, defined as the number of days in the year with minimum temperature below 0°C
- Growing Season Length, defined as the longest consecutive stretch of days in the year with mean temperature above 5°C
- Warm Nights, defined as the number of days in the year with minimum temperature (indicative of nighttime temperature) above the 90th percentile of climatology
- Heat Wave Duration, defined as the longest consecutive stretch of days in the year with maximum temperature exceeding climatological values by more than 5°C
- Extreme Temperature Range, defined as the difference between the warmest daily maximum temperature and the coolest daily minimum temperature in the year

and five indices describing rainfall extreme behavior:

- Consecutive Dry Days, defined as the longest consecutive stretch of days in the year without precipitation
- Precipitation Intensity, defined as the annual average rain amount in wet days
- Number of Days with Rainfall Greater than 10 mm
- Percent of Total Precipitation Falling in Heavy Rain Days, defined as the percent of total yearly precipitation that fell in days whose rain amount exceeded the 95th percentile of wet-day climatology
- 5-Day Maximum Total Precipitation, defined as the largest amount falling in any consecutive 5 days during the year

Nine GCMs computed annual values of these indices from their gridded daily output of temperature (mean, min and max) and precipitation. The annual values were either averaged using the two traditional 20-year windows (present-day, 1980–1999 and future, 2080–2099) and the geographical patterns of the differences analyzed, or low-pass filtered time series (computed as 5-year running means) of global average values were considered.

The behavior of the five indices related to temperature extremes is consistent with what should be expected in a warming world. Heat Waves become longer, Frost Days diminish, Growing Season lengthens, Warm Nights become more numerous. The nine GCMs analyzed agree over the direction of the change, its significance and in large measure also over the geographical patterns of the changes. The analysis looked at three alternative SRES scenarios (high emission, A2, mid-emissions or business as usual, A1B and low-emissions, B1) and found significant differences in the intensification of the warming-related effects between lower and higher emission scenarios, especially in the second part of the twenty-first century. Interestingly however the geographical patterns of changes appear qualitatively similar across scenarios, in agreement with the pattern scaling arguments discussed above. These increases in temperature extremes are mainly the result of higher mean temperatures rather than increased interannual variability, as there is little model agreement on whether temperature variability will change (Meehl et al. 2007b).

Precipitation-related indices present a greater challenge in the quest for model agreement, at least in terms of spatial patterns. There are, however, some general messages that can be gathered from the analysis of the four indices related to intensification of rainfall: there is agreement across models that precipitation intensity will increase almost everywhere over land areas, in larger magnitude in the higher latitudes of the northern hemisphere. The level of inter-model agreement and statistical significance is less uniform over the globe than for temperature-related indices, with patches of regions where changes are not deemed significant by a majority of models. However, when averaged at the global scale all these indices show a significant increase, under all emission scenarios. Consecutive Dry Days is the index with larger inter-annual and inter-model variability. There are nonetheless large areas of the world where changes towards longest dry spells appear with a strong signal, like the Mediterranean basin, central Asia, South Africa, the Amazons and the West and Southwest of the United States.

3.4.3 A Further Look at Growing Season Length

In Tebaldi et al. (2006) growing season was defined in terms of "thermal" characteristics. Obviously, though, moisture and precipitation changes will influence greatly the ability of cultivating crops in areas where structures for irrigation are absent, or water resources are subject to competing demands.

In this section we modify the definition of growing season by including conditions that are related to the available moisture. In addition to requiring mean temperature to be above 5°C we consider a climatology of daily values of the ratio between actual precipitation and potential evapotranspiration. Before taking the ratio, we compute a 10-day moving average of the daily values for each quantity. The growing season starts after the first 5 consecutive days with ratio greater than 0.8 and ends when encountering eight consecutive days with a ratio of less than 0.5 (Thornton et al., 2006). To compute potential evapotranspiration we adopt the formula from Hamon (1961), which uses daily temperature and day length (fraction of hours in the day with sunlight, function of the calendar day and the latitude) as input.

Once the climatological values of the ratio are computed for both present-day climate and future climate, the growing season start- and end-dates are identified and a number of statistics can be computed. Changes in growing season length are of immediate interpretation, but other aspects of a changing climate within the season, like frequency of temperature extremes or dry spells within the growing season, may be evaluated and compared.

Here we comment on some differences between the results based on a thermal definition of growing season length, which suggest a generalized expansion of the growing potential over the calendar year, and the moisture-based definition, which delivers opposite results at least in those regions that are projected to warm the most. In these regions, in the absence of significant changes in precipitation, the increase in potential evapotranspiration causes a moisture deficit that limits the extent of the growing season.

We choose to present some results as area averages over a set of regions of specific relevance for agriculture. They are

- Southern Africa (SAF): 10–30S, 20–35E
- East Africa (EAF): 10S–10N, 30–50E
- Sahel (SAH): 10–15N, 15W–40E
- Northern India (NIN): 25–30N, 70–85E
- East China (ECH): 20–45N, 110–125E
- US Corn Belt (USC): 36–44N, 100–80W
- Western Europe (WEU): 35–50N, 5W–15E
- Australia (AUS): 25–40S, 115–150E

In Table 3.2 we list means (and ranges in parentheses) for projected changes in growing season length under scenario A1B. We list changes by mid-century and end of the century, and for the definition based on evapotranspiration ("ET-based") of growing season, together with the more traditional thermal definition. Units are in number of days.

Table 3.2 Mean number of days (*ranges in parentheses*) for projected changes in growing season length under scenario A1B, using ET and thermal based definitions (see text for details)

Region	ET-based 2046–2065	ET-based 2081–2100	Thermal 2046–2065	Thermal 2081–2100
SAF	−22 (−57,−8)	−29 (−46,−14)	0 (0,1)	0 (0,1)
EAF	2 (−18,18)	4 (−16,25)	0 (0,0)	0 (0,0)
SAH	−10 (−27,12)	−17 (−45,4)	0 (0,0)	0 (0,0)
NIN	5 (−33,41)	11 (−46,62)	9 (3,15)	13 (6,21)
ECH	−5 (−26,23)	−8 (−56,20)	16 (7,23)	22 (8,35)
USC	−4 (−68,41)	−4 (−58,57)	28 (9,62)	41 (18,82)
WEU	−2 (−22,32)	−2 (−30,43)	17 (5,27)	23 (8,37)
AUS	−6 (−43,68)	−8 (−31,15)	0 (0,3)	0 (0,3)

Clearly, including moisture in the definition of growing season changes drastically the nature of future projections for this set of regions. We see a shift from optimistically positive numbers (or no change) across the board to predominantly negative numbers in the mean projections, but with a considerable range of uncertainty spanning both sides of 0 in the individual models' projections. Evidently the effect of temperature is no longer that of simply prolonging mild conditions, conducive to growing crops, but in these instances of exacerbating moisture deficits. The precipitation change signal is not positive enough to balance this off. This is particularly true of areas like Australia, where the distribution of individual model projections lies mostly over negative values by the end of the century, the Sahel and South Africa. All these regions were projected to see either no substantial change or positive changes under the more traditional definition of growing season. Other regions, all seeing an increase under the thermal definitions are now showing large uncertainties, due mainly to a steady increase in average temperatures accompanied by a contradictory set of projections for precipitation change.

Another facet of projected changes potentially affecting the growing season are monthly statistics of extremes of temperature and/or precipitation that may endanger the health of crops in these regions. For example, we may want to extract from the GCM ensemble changes in the number of days when maximum temperature exceeds physiologically critical thresholds (35°C and 40°C) or changes in the average length of dry spells within each month. Figure 3.5 shows an example of this kind of result, for the region of Western Europe, WEU, and three metrics: changes in the number of days with maximum temperature above 35°C, changes in the number of days with maximum temperature above 40°C, and changes in the average length of dry spells. The spaghetti plots show individual model projected changes over the 12 months of the year. There are two sets of lines, dashed and solid, the former showing changes by mid-century, the latter showing changes by end of the century. As we have already pointed out there may be large variability in the numbers projected by each model, but the set of trajectories indicate significant lengthening of dry spells over most of the years, and growing in length the farther in time the projections are. Similarly, large changes in the number of very hot days are projected for the summer months and both time frames.

The results for all the eight regions are described qualitatively in Table 3.3.

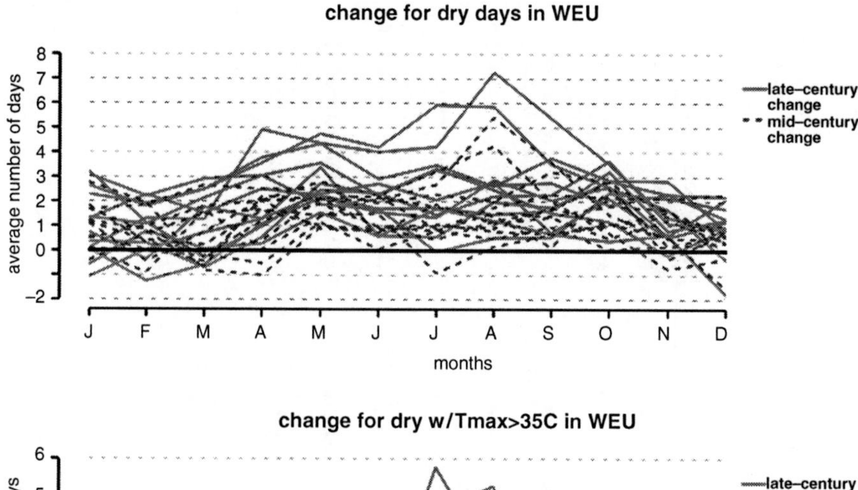

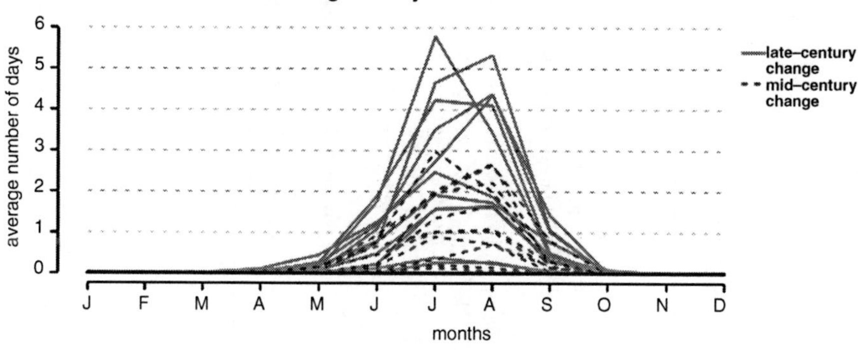

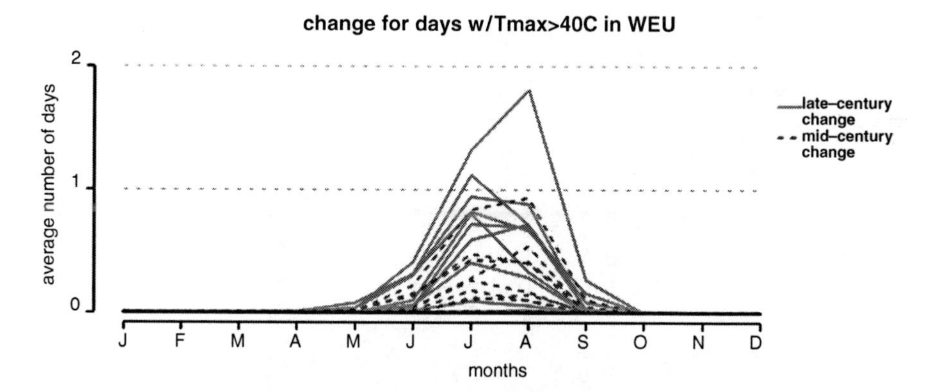

Fig. 3.5 Spaghetti plots for three metrics of extremes relevant to growing season characteristics in Western Europe (WEU). From *top* to *bottom*: changes in the average length of dry spells for each month of the year; changes in the monthly number of days with maximum temperature above 35°C; changes in the monthly number of days with maximum temperature above 40°C. Each line in the plot corresponds to a GCM. *Dashed lines* show changes by mid-century (2046–2065 vs 1981–2000), *solid lines* show changes by end-of-the-century (2081–2100 vs 1981–2000). The scenario is A1B. Thirteen GCMs are represented, all those contributing daily maximum temperature and precipitation to the CMIP3 archive for this scenario

Table 3.3 Summary of changes in growing season characteristics for selected cropping regions

Region	Dry spells length	Max temperature above 35°C	Max temperature above 40°C
EAF	Large uncertainty for most of the year, mostly increase in the summer months	Increase over most of the year	Increase over most of the year
SAF	Some uncertainty in the summer months but mostly increase in the fall and winter months	Increase in the spring and summer months	Increase in the spring and summer months
SAH	Mostly increase in the first six months of the year, larger uncertainty in the latter part	Large increase over most of the year	Large increase over most of the year
NIN	Large uncertainty throughout the May through October months, very small changes during the rest of the year	Increase over most of the year	Increase over most of the year
ECH	Uncertain sign of the change	Increase in the summer months	Increase in the summer months
USC	Uncertain sign of the change	Large increase in the summer months	Large increase in the summer months
WEU	Increase all year long	Increase in the summer months	Increase in the summer months
AUS	Uncertain sign of change in the summer months, increase in the fall and winter months	Increase in the spring and summer months	Increase in the spring and summer months

3.5 Reducing Uncertainties

Uncertainty in future climate stems mainly from scenario and model uncertainty, and both of these for the most part are not intrinsic to the system. In principle, we are free to choose a scenario for future emissions through our actions. Model uncertainty is different, as there is a 'true' climate system, so the uncertainty does not reflect a choice but our incomplete understanding of the system and our inability in describing it in a numerical model. For the decision relevance that climate projections provide, it is also interesting to ask how projection uncertainties will change in the future. If uncertainties are likely to disappear soon, the strategy to wait for better information before spending money may be attractive. If not, then the strategies certainly need to be robust under uncertainty (Lempert and Schlesinger 2000) and an early decision may be wise, for example to have more time to adapt.

Climate models and their projections may improve in the near future in various ways. Some of the errors in the mean state of climate simulations (for example, errors in average temperature and precipitation patterns) appear to develop very fast after the simulations are initiated, suggesting that their causes reside in the behavior of the atmospheric part of the system, rather than the slowly evolving ocean state. Thus, combining weather, seasonal, decadal and long term forecasts in the same modeling framework may improve longer term projections, allowing to develop a better understanding of processes through the verification offered by the shorter term forecasts. This idea of 'seamless prediction' (Palmer et al. 2008) is currently discussed in the scientific community, but it is certainly challenging, both technically and because many assumptions and parameterizations in weather and climate models are not valid across the whole range of spatial and temporal scales that these models would cover. Evaluating models for different climatic states (e.g., the ice age, Otto-Bliesner et al. 2006), variability and trends and on abrupt changes observed in the past can reveal limitations in the model physics.

Short-term predictions may improve through initialization with observations (Smith et al. 2007), assimilation of data or synchronization of multiple models (Kirtman and Shukla 2002). This idea is actually at the basis of the newly developing area of decadal predictions, where a climate model initialized close to the observed state (especially of the oceans, which drive the behavior of the system in the slower frequencies) could be able to generate a climate in sync with the real world, thus moving from projections to actual predictions of the climate system over decadal scales. This new area of research will have to address fundamental questions of predictability (e.g., for how long can we expect two closely initialized versions of the system to stay close?) and methodological issues (e.g., how do we observe enough of the ocean's surface and depth to have a sufficiently accurate representation of the real thing that we want to mimic in our models?). The belief though is that, if successful, these shorter term predictions aiming at simulating not only the overall trend but the decadal oscillations around it would be extremely valuable to impact researchers devising adaption solutions.

Increased computer performance will allow for higher resolution in climate models and more simulations. Resolution will help to improve certain aspects of the simulation (e.g., to resolve topography, or convection) but it does not necessarily help in the case where the processes are poorly understood (e.g., how to parameterize the effect of vegetation on climate and the water cycle). More simulations will be useful to better quantify the uncertainty in the models. Better observations, in particular long-term records of observed changes, will further constrain the models and help to understand processes critical to improving model performance.

Many important quantities in the climate system are only observed since the advent of satellites, making it difficult to separate long-term trends from natural variability. A hierarchy of models (Held 2005) with different structures and families of similar models can be used to track a behavior across models, and to identify which quantities are most useful to identify model deficiencies and to constrain future projections (e.g., Hall and Qu 2006). Parametric uncertainty in a model can possibly be reduced by calibration if computational capacity is large enough to run the model many times. The structural uncertainty, i.e. the fact that the model is unable to match all observations for any set of parameters, or that different model formulations may do similarly well at reproducing observations and cannot be distinguished, is harder to eliminate.

Given the complexity of the system we are trying to describe and predict, and the large number of processes, interactions and feedbacks occurring on different spatial and temporal scales, the uncertainty in climate projections may not decrease quickly in the near future. The present climate seems to provide only a weak constraint on the future, and models continue to improve in simulating the present (Reichler and Kim 2008) but they do not clearly converge on the future trends. Models also continue to include more processes and feedbacks interactively, giving rise to new sources of uncertainty and hitherto unknown interactions Some uncertainties are intrinsic and irreducible (e.g., the chaotic nature of short-term weather and climate variability which limits the predictability on timescales of weeks to years, or the timing of volcanic eruptions in the future).

It is therefore important that scientists specify all possible outcomes, rather than trying to reduce spread where it cannot be reduced. The situation is difficult in that overly optimistic and tight uncertainties may make society vulnerable if things turn out to be different from what was predicted. On the other hand, providing large uncertainty estimates can prevent action, and is often seen as being alarmist because extreme changes are not ruled out. Some constraints will emerge as climate change proceeds, so even with the same models and methods, we expect uncertainties to shrink somewhat. In some situations they may also grow, if the additional data reveal that the model is imperfect, and that further processes need to be considered.

References

Barnett T, Zwiers FW, Hegerl GC et al. (IDAG) (2005) Detecting and attributing external influences on the climate system: a review of recent advances. J Clim 18:1291–1314

Bony SR, Colman VM, Kattsov RP, et al. How well do we understand and evaluate climate change feedback processes? J Climate. 2006;19:3445–3482.

Christensen JH, Christensen OB. A summary of the PRUDENCE model projections of changes in European climate by the end of this century. Climatic Change. 2007;81(1):7–30.

Christensen JH, Hewitson B, Busuioc A, et al. Regional climate projections. In: Solomon S, Qin D, Manning M, et al., editors. The physical science basis. Contribution of working Group I to the fourth assessment report of the intergovernmental panel on climate change. Cambridge, United Kingdom and New York, NY, USA: Cambridge University Press; 2007.

Claussen M, Mysak LA, Weaver A, et al. Earth system models of intermediate complexity: closing the gap in the spectrum of climate system models. Climate Dynamics. 2002;18(7):579–586.

Friedlingstein P, Cox P, Betts R, et al. Climate-carbon cycle feedback analysis: results from the C4MIP model intercomparison. J Climate. 2006;19:3337–3353.

Giorgi F, Francisco R. Uncertainties in regional climate change predictions. A regional analysis of ensemble simulations with the HADCM2 GCM. Climate Dynamics. 2000;16:169–182.

Giorgi F, Mearns LO. Calculation of average, uncertainty range and reliability of regional climate changes from AOGCM simulations via the 'reliabil-ity ensemble averaging' (REA) method. J Climate. 2002;15:1141–1158.

Giorgi F, Mearns LO. Probability of regional climate change calculated using the reliability ensemble average (REA) method. Geophys Res Lett. 2003;30(12):1629–1632.

Gleckler PJ, Taylor KE, Doutriaux C. Performance metrics for climate models. J Geophys Res. 2008;. doi:10.1029/2007JD008972.

Hall A, Qu X. Using the current seasonal cycle to constrain snow albedo feedback in future climate change. Geophys Res Lett. 2006;. doi:10.1029/2005GL025127.

Hamon WR (1961) Estimating potential evapotranspiration. J Hydraul Div 87(HY3):107–120

Hegerl GC, Zwiers FW, Braconnot P, et al. Understanding and attributing climate change. In: Solomon S, Qin D, Manning M, et al., editors. Climate change 2007: the physical science basis. Contribution of working group I to the fourth assessment report of the intergovernmental panel on climate change. Cambridge, United Kingdom and New York, NY, USA: Cambridge University Press; 2007.

Held IM. The gap between simulation and understanding in climate modeling. Bull Am Meteorol Soc. 2005;86(11):1609–1614.

Kirtman BP, Shukla J. Interactive coupled ensemble: a new coupling strategy for GCMs. Geophys Res Lett. 2002;29:1029–1032.

Knutti R, Hegerl GC. The equilibrium sensitivity of the Earth's temperature to radiation changes. Nature Geoscience. 2008;1:735–743.

Knutti R, Stocker TF, Joos F, Plattner GK. Constraints on radiative forcing and future climate change from observations and climate model ensembles. Nature. 2002;416:719–723.

Knutti R, Allen MR, Friedlingstein P, et al. A review of uncertainties in global temperature projections over the twenty-first century. J Climate. 2008;21:2651–2663.

Lempert RJ, Schlesinger ME. Robust strategies for abating climate change. Climatic Change. 2000;45(3–4):387–401.

Lobell DB, Field CB. Global scale climate–crop yield relationships and the impacts of recent warming. Environ Res Lett. 2007;. doi:10.1088/17489326/2/1/014002.

Lobell DB, Burke MB, Tebaldi C, Mastrandrea MD, Falcon WP, Naylor RL. Prioritizing climate change adaptation needs for food security in 2030. Science. 2008;319(5863):607–610.

Meehl GA, Covey C, Delworth T, et al. The WCRP CMIP3 multimodel dataset. Bull Am Meteorol Soc. 2007a;88:1383–1394.

Meehl GA, Stocker TF, Collins WD, et al. Global climate projections. In: Solomon S, Qin D, Manning M, et al., editors. The physical science basis. Contribution of working Group I to the fourth assessment report of the intergovernmental panel on climate change. Cambridge, United Kingdom and New York, NY, USA: Cambridge University Press; 2007b.

Murphy JM, Sexton DMH, Barnett DN, Jones GS, Webb MJ, Collins M, et al. Quantification of modeling uncertainties in a large ensemble of climate change simulations. Nature. 2004;430:768–772.

Nakicenovic N, Swart R. IPCC special report on emissions scenarios. Cambridge, United Kingdom and New York, NY, USA: Cambridge University Press; 2000.

Otto-Bliesner BL, Brady E, Clauzet G, et al. Last glacial maximum and holocene climate in CCSM3. J Climate. 2006;19:2526–2544.

Palmer TN, Doblas-Reyes FJ, Weisheimer A, Rodwell MJ. Toward seamless prediction calibration of climate change projections using seasonal forecasts. Bull Am Meteorol Soc. 2008;89(4):459–470.

Räisänen J. How reliable are climate models? Tellus. 2007;59(1):2–29.

Räisänen J, Palmer TN. A probability and decision-model analysis of a multimodel ensemble of climate change simulations. J Climate. 2001;14:3212–3226.

Randall DA, Wood RA, Bony S, et al. Climate models and their evaluation. In: Solomon S, Qin D, Manning M, et al., editors. The physical science basis. Contribution of working group I to the fourth assessment report of the intergovernmental panel on climate change. Cambridge, United Kingdom and New York, NY, USA: Cambridge University Press; 2007.

Reichler T, Kim J. How well do coupled models simulate today's climate. Bull Am Meteorol Soc. 2008;89(3):303–312.

Santer BD, Wigley TML, Schlesinger ME, Mitchell JFB. Developing climate scenarios from equilibrium gcm results. Technical note 47. Hamburg: Max Planck Institut für Meteorologie; 1990. p. 29.

Smith DM, Cusack S, Colman AW, et al. Improved surface temperature prediction for the coming decade from a global climate model. Science. 2007;. doi:10.1126/science.1139540.

Smith RL, Tebaldi C, Nychka DW, Mearns LO. Bayesian modeling of uncertainty in ensembles of climate models. J Am Stat Ass. 2009;104(485):97–116.

Solomon S, Qin D, Manning M, et al., editors. Climate change 2007: the physical science basis. Contribution of working group I to the fourth assessment report of the intergovernmental panel on climate change. Cambridge, United Kingdom and New York, NY, USA: Cambridge University Press; 2007.

Stainforth DA, Aina T, Christensen C, et al. Uncertainty in predictions of the climate response to rising levels of greenhouse gases. Nature. 2005;433:403–406.

Stott PA, Kettleborough JA. Origins and estimates of uncertainty in predictions of twenty-first century temperature rise. Nature. 2002;416:723–726.

Tebaldi C, Knutti R. The use of the multi-model ensemble in probabilistic climate projections. Philos Trans R Soc A. 2007;1857:2053–2075.

Tebaldi C, Lobell DB. Towards probabilistic projections of climate change impacts on global crop yields. Geophys Res Lett. 2008;. doi:10.1029/2008GL033423.

Tebaldi C, Mearns LO, Nychka DW, Smith RL. Regional probabilities of precipitation change: a Bayesian analysis of multimodel simulations. Geophys Res Lett. 2004;. doi:10.1029/2004GL021276.

Tebaldi C, Smith RL, Nychka DW, Mearns LO. Quantifying uncertainty in projections of regional climate change: a Bayesian approach to the analysis of multi-model ensembles. J Climate. 2005;18(10):1524–1540.

Tebaldi C, Hayhoe K, Arblaster JM, Meehl GA. Going to the extremes: an intercomparison of model-simulated historical and future changes in extreme events. Climatic Change. 2006;79:185–211.

Thornton PK, Jones PG, Owiyo T, Kruska RL, Herrero M, Kristjanson P, Notenbaert A, Bekele N and Omolo A, with contributions from Orindi V, Otiende B, Ochieng A, Bhadwal S, Anantram K, Nair S, Kumar V, Kulkar U (2006) Mapping climate vulnerability and poverty in Africa. Report to the Department for International Development, ILRI, PO Box 30709, Nairobi 00100, Kenya, pp 171

Vrac M, Stein ML, Hayhoe K, Liang XZ. A general method for validating statistical downscaling methods under future climate change. Geophys Res Lett. 2007;. doi:10.1029/2007GL030295.

Part II

Chapter 4
Crop Response to Climate: Ecophysiological Models

Jeffrey W. White and Gerrit Hoogenboom

Abstract To predict the possible impacts of global warming and increased CO_2 on agriculture, scientists use computer-based models that attempt to quantify the best-available knowledge on plant physiology, agronomy, soil science and meteorology in order to predict how a plant will grow under specific environmental conditions. The chapter reviews the basic features of crop models with emphasis on physiological responses to temperature and CO_2 and explains how models are used to predict potential impacts of climate change, including options for adaptation. The closing section reviews major issues affecting the reliability of model-based predictions. These include the need for accurate inputs, the challenges of improving the underlying physiological knowledge, and the need to improve representations of genetic variation that likely will affect adaptation to climate change.

4.1 Introduction

Ecophysiological models were the dominant tools used to estimate the potential impact of climate change in agroecosystems in the Third and Fourth Assessment Reports of the IPCC (Gitay et al. 2001; Easterling et al. 2007) and are widely used elsewhere in climate change research. These models, also known as "crop models" or "simulation models", attempt to encapsulate the best-available knowledge on plant physiology, agronomy, soil science and agrometeorology in order to predict how a

J.W. White (✉)
US Arid Land Agricultural Research Center, USDA-ARS, 21881 North Cardon Lane, Maricopa, AZ 85224
email: jeffrey.white@ars.usda.gov

G. Hoogenboom
Biological and Agricultural Engineering, 165 Gordon Futral Court, The University of Georgia, Griffin, GA 30223-1731
email: Gerrit@uga.edu

D. Lobell and M. Burke (eds.), *Climate Change and Food Security*,
Advances in Global Change Research 37, DOI 10.1007/978-90-481-2953-9_4,

plant will grow under specific environmental conditions. The models are "ecophysiological" because they use mathematical descriptions of physiological, chemical and physical processes to simulate crop growth and development over time. Physiological processes considered may include photosynthesis, respiration, growth and partitioning, development of reproductive structures, transpiration, and uptake of water and nutrients. Chemical and physical processes can involve soil chemical transformations, energy flows, and diffusion of gases into and out of leaves, among others.

To predict crop growth, the model requires that initial conditions be specified, such as the soil nutrient and water status, the planting date and density. Data on temperature, solar radiation, precipitation, or other weather parameters are then used to estimate how the development and growth of the crop progress over the cropping season. Most models operate at daily time steps, starting at planting and ending at the prediction of harvest or physiological maturity, depending on the crop. Information on irrigations, fertilizer applications, tillage events, pests, diseases, or other factors also may be considered.

The first ecophysiological models were developed by De Wit (1965) in the Netherlands and Duncan and colleagues in the United States (Duncan et al. 1967). These models were primarily used as research platforms to quantitatively test basic hypotheses about plant growth and development (Loomis et al. 1979). As computing power increased and understanding of basic processes improved, more factors were considered, such as the dynamics of specific nutrients in the soil and plant and the effects of pests and diseases.

Models are available for all major annual crops and many minor crops (e.g., Jones et al. 2003). Current models typically run on a personal computer or a work station and can simulate a cropping season in less than 1 s. The responsiveness of the models to climate and other environmental variables, as well as to crop management, allow them to be easily adapted to simulate responses to projected climate conditions, such as obtained from general circulation models.

This chapter first reviews how ecophysiological models function with emphasis on the physiological responses that are most relevant to climate change research. We then discuss how the models are applied in climate change research and identify challenges and opportunities for improving the models per se and how they are applied to climate change research. The overall objective is to help readers understand how the features of ecophysiological models affect projections of potential impacts of climate change. Readers seeking further information should consult texts such as Hay and Porter (2006) for an overview of the physiological assumptions and Tsuji et al. (1998) or Hanks and Ritchie (1991) for details on modeling both plant and soil processes.

4.2 Overview of Ecophysiological Models

A simulation using a basic model might start with a set of initial conditions specifying where the crop is grown, the initial status of water and nutrients in the soils, and the parameters needed to represent the physiological characteristics of the crop.

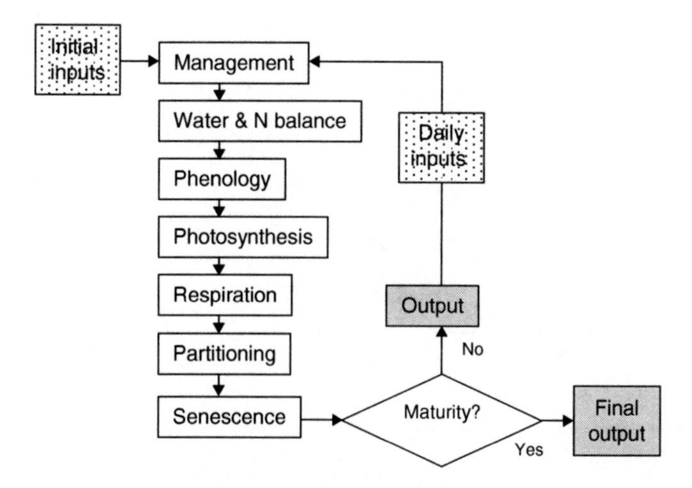

Fig. 4.1 Flow diagram for a hypothetical ecophysiological model with a daily time step

For an annual crop, the model loops through a series of subroutines that estimate plant or soil processes on an hourly or daily basis, outputting intermediate values at specified intervals (Fig. 4.1). In each cycle, the model checks whether the crop has reached maturity or a harvest data, in which case the yield and a diverse range of summary data may be output. Normally, the output of these models consists of a series of continuous curves representing different plant or environmental variables that change over time. Figure 4.2 presents examples of such output from the Cropping System Model (CSM)-CROPGRO model for a single crop of common bean grown near Cali, Colombia.

Mathematically, a model integrates a system of differential equations that describe rates that vary over time. The predicted (state) variables for the crop may include the dry mass of organs, leaf area, root length and vertical distribution in the soil, developmental progress, and soil water and nutrient concentrations of individual soil layers or horizons. In practice, the equations are far too complex for analytical solutions, so they are integrated numerically using time steps of a few seconds in very detailed models or hourly to daily, as found in most models.

Hundreds of ecophysiological models have been created. Many of these were developed by either a single scientist or small teams for a single research purpose. Most of these models can now only be found in the literature, although their algorithms may persist in newer models. There is no formal system of nomenclature, and in some cases, a single model has been modified independently by different groups, resulting in confusion over versions referred to in publications. Table 4.1 lists four families of models that have seen widespread use in climate change research.

The basic processes represented in ecophysiological models are described here mainly with reference to a hypothetical average plant. Most models actually report outputs on a land area basis, which corresponds to a community of identical "average" plants. A few models can simulate genetic mixtures, either of the same species or different species, including weeds.

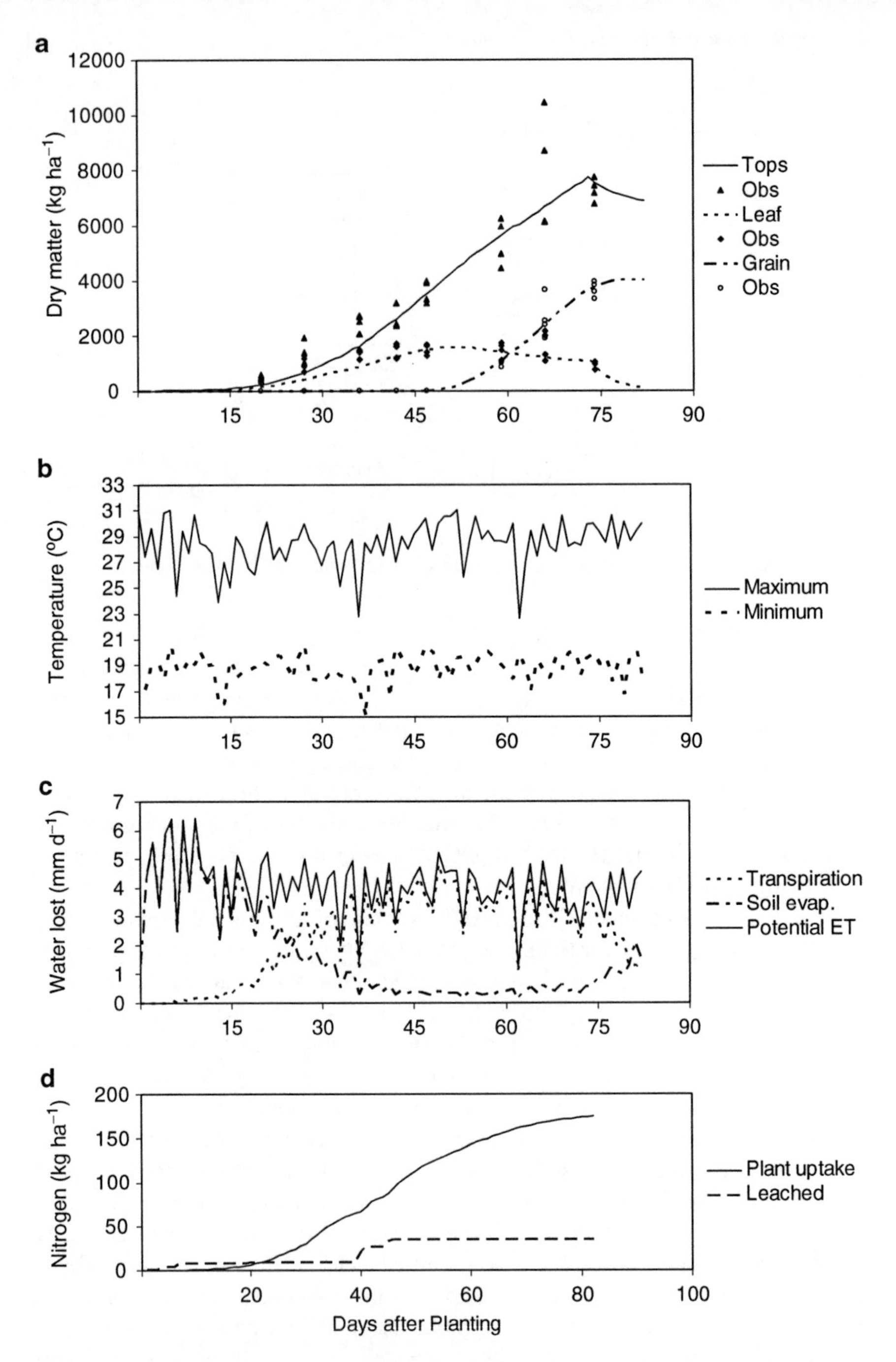

Fig. 4.2 Simulation of growth of common bean at Palmira, Colombia using the CSM-CROPGRO-Drybean model. (**a**) Change in dry matter for tops (total aboveground material), leaves and grains. Symbols indicate observed data from individual replicates. (**b**) Daily maximum and minimum temperatures. (**c**) Transpiration, evaporation from soil, and potential evapotranspiration. (**d**) Nitrogen taken up by plants or lost through leaching

Table 4.1 Examples of ecophysiological models used in climate change research

Name	Crop species	Description and references	Example applications
CropSyst	Barley, maize, sorghum, soybean, wheat, and others	Radiation use efficiency. Daily time step. Allows cropping sequences (Stockle et al. 2003)	• Rainfed wheat in south-eastern Australia (Anwar et al. 2007) • Bambara groundnut, peanut, maize, sorghum and soybean in Cameroon (Tingem et al. 2008)
CSM-CERES	Barley, maize, millet, sorghum, and wheat	Radiation use efficiency. Daily time step. Allows cropping sequences (Jones et al. 2003; Hoogenboom et al. 2004)	• Maize and winter wheat in the United States (Alexandrov and Hoogenboom 2000) • Wheat and maize in the Iberian Peninsula (Minguez et al. 2007) • Maize production in Africa and Latin America (Jones and Thornton 2003)
CSM-CROPGRO	Common bean, faba bean, peanut, soybean and other legumes, cotton, and others	Farquhar-type photosynthesis calculated on an hourly basis with a hedge-row model for light interception. Considers growth and maintenance respiration. Allows cropping sequences (Jones et al. 2003; Hoogenboom et al. 2004)	• Soybean and peanut in the United Sates (Alexandrov and Hoogenboom 2000) • Soybean in northeastern Austria (Alexandrov et al. 2002) • Soybean in southern Québec, Canada (Brassard and Singh, 2008)
EPIC	Maize, millet, rice, sorghum, soybean, wheat, and others	Radiation use efficiency. Daily time step. Allows cropping sequences and can model effects of tillage and soil erosion (Williams et al. 1989)	• Maize, soybean and wheat in the Midwestern US, considering different spatial scales for climate and soil (Easterling et al. 2001) • Maize, sorghum, millet, rice and cassava in Nigeria (Adejuwon 2006)

4.2.1 Development

Development includes the processes used by a plant to schedule important changes in growth such as the seedling emergence, formation of flowers, the onset of rapid grain growth, or the end of grain growth, which usually is considered to represent physiological maturity. This life history can be interpreted as a series of phases that are demarcated by stages, so the modeling approach is often termed "phasic development" (Ritchie and NeSmith 1991). Each phase is characterized by a duration that is expressed in physiological time, which is mathematically similar to thermal time, growing degree days or heat units but may include influences of photoperiod, vernalization or other processes. The duration represents the minimum time required for the plant to progress from one stage to another under optimum conditions. Each day (or hour), the plant is assumed to progress in time at a developmental rate (D_t), which is estimated from a potential rate (D_p) and potential rate adjusting factors such as for temperature (T), photoperiod (P) and water deficits (W):

$$D_t = D_P * T * P * W$$

The rate adjusting factors usually vary from 0 to 1 in order to slow development below the maximum rate, but stresses such as water deficits may be used to accelerate development, resulting in the factor exceeding a value of 1. An alternate approach to phasic development that is especially common in modeling cereals is to use leaf number as the main indicator of developmental progress. While the terminology differs, the underlying physiology is similar (Jamieson et al. 2007).

The details of how phenology is modeled differ greatly with the biology of the crop species and decisions of the model developers concerning how to represent specific responses. For temperature, the decisions involve the selection of the temperature variables and specification of a curve that describes the assumed shape of a given physiological response. Common assumptions are that each crop has a "base temperature", below which it does not grow or develop and an "optimum temperature" that allows the maximum rate of growth or development. Temperatures may be observed or estimated from hourly values, daily averages, or averages adjusted or weighted in various manners. Of course, a crop does not respond to a daily value of the maximum or minimum temperatures; it is exposed to temperature (and all other environmental conditions) on a continuous basis. The models use simplified temperature data and associated equations. Soil temperature is often used to control germination, seedling emergence, and in cereals, early development of the shoot since the crown remains close to the soil surface.

Response curves vary from simple "broken stick" models to non-linear functions such as the beta function. Cardinal temperatures identify transition points in these responses. Besides the base and optimal temperatures described previously, models differ in how effects of supra-optimal temperatures are represented. The simplest approach is to assume that the maximum developmental rate is sustained above the designated optimum. Alternately, the rate may be assumed to decline to a lethal temperature, considered the maximum temperature for development, or the maximal

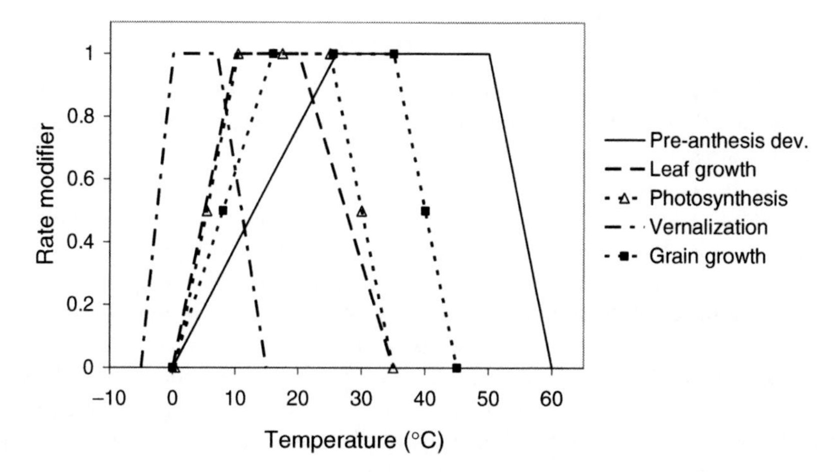

Fig. 4.3 Examples of temperature response functions assumed in the CSM-Cropsim-CERES wheat model. Curves are for pre-anthesis development, leaf growth, photosynthesis, vernalization, and grain growth. All responses are based on daily mean temperature

rate may be sustained up to a second optimum above which the rate decreases to the lethal temperature (Fig. 4.3). This is an area that requires further research, especially as it relates to projected increases in temperature.

To non-specialists, the diverse approaches for modeling development may seem unscientific. The causes of the diversity are complex and reflect fundamental difficulties in simulating plant responses to weather conditions in general. Foremost is that, while air or soil temperatures can be measured accurately, a plant in a community experiences a complex, fluctuating temperature environment. Temperature sensing for a given process may reside in a specific tissue, such as the shoot apical meristem for vernalization (Sung and Amasino 2004). Models often assume that the air temperature reported from the nearest weather station approximates an average above-ground crop temperature, but there often are large temperature gradients within a canopy (Desjardins et al. 1978). Another challenge is that temperature responses involve circadian rhythms (the internal biological clocks) of the plant, and results from studies under constant temperatures or from simple constant day/night regimes likely have limited utility for quantifying temperature responses. A further problem is that response variables such as time of floral initiation or onset of flowering are usually scored visually on a sample of plants that vary in their developmental progress. These scores have error due to observation bias and sampling.

4.2.2 Growth

Growth is described through accumulation of dry matter in the main plant organs plus changes in a few additional traits such as leaf area and root length. An assimilate balance for a given time interval may be expressed as

$$G = P - (R + S)$$

where G is a growth increment per unit time, P is net photosynthesis for the plant or crop, and R and S are losses due to respiration and senescence (death of tissues related to stress or aging).

At the single leaf level, photosynthesis is simulated in response to light intensity, temperature and leaf external CO_2 concentration. The Farquhar, von Caemmerer and Berry model for leaf photosynthesis (Farquhar et al. 1980) is often used for species with the C_3 photosynthetic pathway, and the basic model is readily extended to account for the concentration of CO_2 for the C_4 pathway (von Caemmerer, 2000).

Temperature and $[CO_2]$ are obtained as external inputs or from other routines of the model. Estimating the light level, or more specifically the photon flux density for photosynthetically active radiation, requires describing how radiation is intercepted by the canopy. Many simple approaches assume that the irradiance (I) declines exponentially with the leaf area index (leaf area per unit land area, L),

$$I = I_0 * e^{-K*L}$$

where K is a dimensionless extinction coefficient that varies from 0 to 1. A canopy that predominantly contains horizontally oriented leaves has a higher value of K, and I declines more rapidly. Numerous complications are introduced when consideration is given to the diurnal cycle of radiation, effects of canopy shape and leaf angle distribution, diffuse and direct components of radiation, reflection from leaves, and other factors (Hay and Porter 2006). Pursuing these complications, however, may bring little benefit in accuracy where solar radiation data are unavailable and have to be estimated. We note especially that estimation of changes in solar radiation with climate change remain problematic for global circulation models (GCMs).

Plant tissues that are not actively photosynthesizing release CO_2 through respiration just like any heterotrophic organism. This is because metabolic activity requires energy, whether it is to maintain existing tissue, construct new tissue, take up nutrients, or transport sugars. Models typically recognize two components to respiration. Growth respiration occurs in the construction of new tissues. Its rate varies primarily with the composition of the tissues being synthesized because the metabolic cost of synthesizing lipid, protein or lignin is much higher than for cellulose or starch (Penning de Vries et al. 1974). Maintenance respiration involves transport of nutrients, protein turnover, maintenance of ion gradients across membranes, and a host of other processes that are difficult to monitor individually. This component is usually assumed to increase with temperature and plant protein content, which is a good indicator of the overall metabolic activity of the plant and is proportional to total plant biomass.

Senescence is the process of controlled death of tissues. Leaf death is the most readily observed form, but stems, roots and fruits also senesce. Leaf senescence is largely associated with either shading or aging of early-formed leaves as the canopy develops or with mobilization of nitrogen during grain filling. Other drivers of senescence include water deficits, heat stress, flooding, and cold or frost damage. Typically, a moderate stress slows growth but if a threshold is exceeded, senescence occurs.

A widely used alternative to simulating dry matter growth through component processes is to evaluate G on a daily basis by assuming that net daily growth is the product of light intercepted by the canopy (I) and an integrative conversion factor called radiation use efficiency (RUE):

$$G = I * RUE$$

As with other estimators of photosynthesis, RUE can be modeled using a potential or reference value that varies with genotype, temperature, atmospheric CO_2 or specific environmental stresses.

A comparison of the CSM-CERES models for maize, rice and sorghum illustrates the complexity underlying seemingly simple approaches (Fig. 4.4). Firstly, the temperature responses are based on a weighted average, TAVGD, calculated from the daily maximum (TMAX) and minimum (TMIN) as

$$TAVGD = 0.25 * TMIN + 0.75 * TMAX.$$

The averaging implies that daytime temperatures have a greater effect on the processes underlying G than night temperatures. In comparing the respective rate modifiers for maize, rice and sorghum (Fig. 4.4), the responses for maize and rice are similar, not withstanding that C_4 species such as maize and sorghum are generally considered more heat tolerant than C_3 species like rice. The sorghum response agrees with the expectation that this species is more heat tolerant than maize. The curves, however, only partially define the response of G to temperature because the

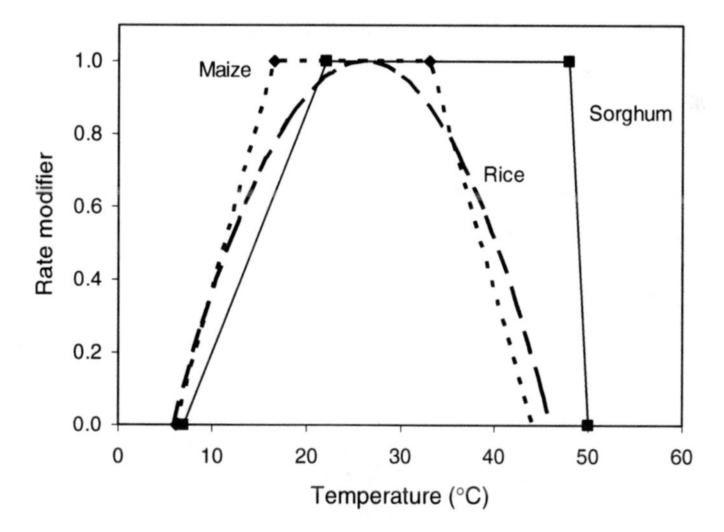

Fig. 4.4 Examples of temperature response functions assumed in the CSM-CERES models for radiation use efficiency of maize, rice and sorghum. All responses are based on a weighted average temperature calculated as 0.75 of the daily maximum and 0.25 of the minimum

models assume that when temperature, nitrogen, water, or other stresses affect G, only the most severe stress is effective.

The [CO_2] response for the CSM-CERES models is applied regardless of impacts of other environmental factors. The reference values of RUE used by the models are assumed to have been estimated for recent historic conditions with [CO_2] of 350 ppm, so values above this level increase growth rate (Fig. 4.5). The expected greater responsiveness of species with C_3 photosynthesis is in accordance with the basic expectations. However, the use of only two curves reflects the scarcity of reliable data on field-level responses to [CO_2] rather than a consensus that there are no differences among the species (see Chapter 7).

Leaf area growth is modeled in order to estimate light interception. A common approach is to estimate an increment in leaf area from new leaf mass using the leaf area to mass ratio, also known as the specific leaf area (SLA). A reference value of SLA may be input as a cultivar specific parameter, and the actual SLA applied for a growth increment is varied with crop physiological age, temperature, solar radiation or other factors.

Simulating water or nutrient uptake requires information on the distribution of roots in the soil, including root length. Once a root mass increment is determined, root length growth varies with the tendency of the crop to be deep or shallow-rooted, the length to mass ratio of new roots, the current root length distribution, and soil physical constraints. Downward growth of roots is mainly temperature driven. Jones et al. (1991) reviewed these processes in more detail.

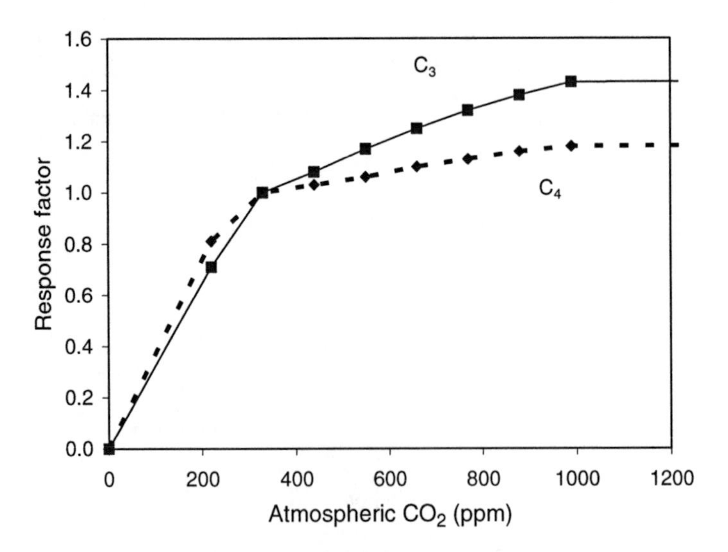

Fig. 4.5 Response to atmospheric [CO_2] assumed in the CSM-CERES models for radiation use efficiency of C_3 crops (barley, oats, rice and wheat) and C_4 crops (maize and sorghum)

4.2.3 *Partitioning*

The simulated increase in dry matter or growth is allocated to different plant organs by following a set of rules for partitioning that change with development and sometimes also with different stress levels. In early vegetative growth, priority often goes to leaf growth, with an additional portion for stems to ensure that there is supporting structure for the leaves. Root growth is adjusted to ensure that enough roots are formed to satisfy demand for water or specific nutrients. As the plant grows, it may produce more assimilates than are required to sustain the estimated maximum possible growth rate of organs. This excess may be stored in roots or shoots, or it may be allocated to a pool of remobilizable reserves. As the plant transitions to reproductive growth, partitioning to reproductive structures begins. The details of how partitioning to reproductive growth is simulated vary with the morphology of the crop and the physiological complexity of the model. Simple models apportion assimilate to reproductive organs by assuming a linear increase in harvest index over the grain-filling phase (Spaeth and Sinclair 1985). Alternately, growth of individual fruits or cohorts of fruits is simulated, allowing for competition among the fruits to senesce the least competitive fruits (Boote et al. 2002). Further complications arise in attempting to simulate growth of individual seeds, especially if protein or oil contents are considered. The final seed yield is variously determined at physiological maturity, which is estimated from routines for development, or at the harvest date, which may be estimated or provided as an input.

Partitioning rules for tuber and root crops are similar except that in place of flowering or onset of grain filling, a stage of onset of tuber or storage root growth demarcates major changes in partitioning rules (Singh et al. 1998). In such models, growth of storage organs is described separately from fibrous root growth.

4.2.4 *Environment*

Environmental factors such as temperature, solar radiation, and [CO_2] directly affect plant processes, and models can use the current value of a factor, such as from a daily weather record, with minimal modification. The effects of these factors have been discussed individually in relation to specific processes, but their roles are summarized in Table 4.2. It is constructive to compare responses across processes.

4.2.4.1 Temperature

The CSM-Cropsim-CERES-Wheat model specifies nine temperature responses affecting development, photosynthesis, and different aspects of growth. Three responses for grain formation and growth have identical responses, so in practice,

Table 4.2 Summary of the effects of selected environmental factors on plant processes, including a subjective assessment of how completely current models simulate a given process

Factor	Process	Modeled	Comments
Temperature	Phenology	Full	Warming usually decreases time to flowering and maturity, but high temperatures may delay development
	Photosynthesis	Full	Heat stress is poorly understood and seldom explicitly modeled
	Respiration	Full	Rates increase with temperature
	Leaf development	Partial	Models differ greatly how temperature affects leaf expansion and thickness
	Reproductive growth	Partial	Heat stress is poorly understood and seldom explicitly modeled
	Root elongation	Full	Rate increases with soil temperature, but soil temperatures are poorly modeled, including for climate change conditions
	Potential evapotranspiration	Full	Potential water loss increases with temperature, as accurately predicted by Penman–Monteith equation
	Mineralization of soil organic matter	Full	Rates increase with soil temperature
CO_2 concentration	Development	Not	Effects vary with species and are not well enough understood to be modeled
	Leaf development	Not	Not well enough understood to be modeled
	Photosynthesis	Full	Basic response to CO_2 is well described by the Farquhar model, but controversies remain
	Respiration	Not	Not fully accepted as existing
	Transpiration	Full	Physiological mechanisms are poorly understood. Cultivar differences are likely but not considered in current models
Solar radiation	Photosynthesis	Full	Leaf and canopy responses are well described by models
	Leaf development	Partial	Few models consider effects on leaf expansion and thickening
	Potential evapotranspiration	Full	Potential water loss increases with radiation, as accurately predicted by Penman–Monteith equation
Wind	Potential evapotranspiration	Partial	Potential water loss increases with wind, as accurately predicted by Penman–Monteith equation
Relative humidity	Leaf development	Not	Not well enough understood to be modeled
	Potential evapotranspiration	Partial	Potential water loss decreases with humidity, as accurately predicted by Penman–Monteith equation
	Transpiration	Partial	Direct plant responses to humidity, including cultivar differences, are poorly understood

seven unique responses are recognized. Each response is described with four cardinal temperatures using a trapezoidal response curve (Fig. 4.3). Vernalization (the requirement some crops exhibit for a cold period) is unique in that it only operates from −5°C to 10°C, reflecting that this process quantifies a specific low temperature response. Pre-anthesis development is shown as continuing up to 60°C, but since average temperatures never reach these levels, the real and practical result is that the development rate is maximal above 25°C.

The actual curves are estimated through diverse procedures. Cardinal temperatures for development can be estimated with non-linear optimization, using field or controlled environments as data sources (e.g., Grimm et al. 1993). Specific physiological responses may be estimated by compiling data across studies. To define a response of leaf photosynthesis to temperature in wheat, Bindraban (1999) examined data from six publications, the previously established response for SUCROS, and his own field measurements. The review of temperature responses for wheat by Porter and Gawith (1999) shows the diversity of values that may be found for a single crop.

When multiple temperature responses are applied in real world situations, the crop responses can be surprisingly complex. For example, to assess the impact of a temperature increase on irrigated sorghum production in Arizona, one might examine the base response of contrasting hybrids to planting date using historic weather data and a simple increase of +1.5°C for daytime temperatures and +3.0°C for the nighttime (Fig. 4.6). For both temperature regimes, the hybrid Cargill 877 is about 20 days later than Cargill 577 for planting dates up to mid-August, when the difference increases due to cooler temperatures (Fig. 4.6a). The warmer temperature regime accelerates development resulting in earlier flowering, although the difference is less than 5 days for plantings from April through August. The response of grain yield (Fig. 4.6b), however, is much more complex. Yields are similar for both hybrids and climate regimes with mid-February plantings, but by early June, warming is predicted to reduce yields by about 500 kg ha^{-1}. The yield effect increases up to August when there is a change in response; the warming regime becomes advantageous relative to the historic regime because the warming extends the growing season.

4.2.4.2 Water

Water, nitrogen, and other factors that involve uptake from the soil into the plant require consideration of the availability of the resource in the soil, demand by the plant for the resource, and the ability of the plant to take up the resource via the roots. Consideration is also required of alternate pathways such as evaporation of water from the soil surface and return of nitrogen to the soil from senesced and abscised leaves. For a given soil resource, the overall basic process is readily described with an equation that balances sources of the resource against losses.

For water, sources may be precipitation (P) and irrigation (I), and losses are through evaporation (E), transpiration (T), surface runoff (R), storage in the soil (S), and deep percolation (D). Thus,

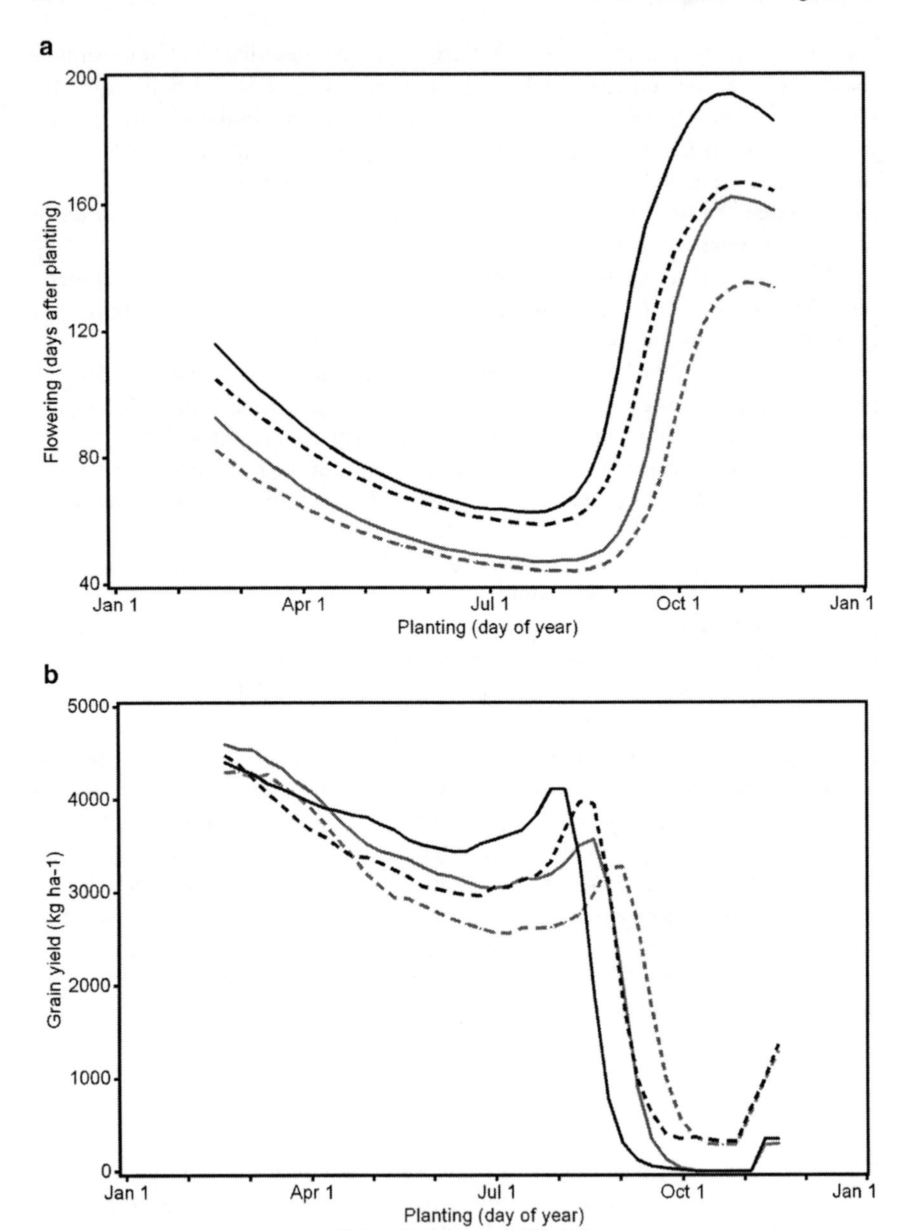

Fig. 4.6 Simulations using the CSM-CERES Sorghum model for response of planting date in two sorghum hybrids, Cargill 577 and Cargill 877, at Maricopa, Arizona. Weather data are for 14 years of historic conditions and for the same years assuming a warming scenario of +1.5°C for the daily maximum and +3.0°C for the minimum. (**a**) Days to flowering. (**b**) Grain yield

$$(P + I) = (E + T + R + S + D)$$

The values of P, I and R are essentially predetermined.

To estimate E and T, models first calculate the potential atmospheric demand through evaporation and transpiration, termed potential evapotranspiration (PET), from weather variables and crop canopy conditions. PET increases with solar radiation, temperature, and wind but decreases with relative humidity. The adaptation of the Penman–Monteith equation for PET (Monteith and Unsworth 1990) that is described in the FAO Drainage and Irrigation Paper No. 56 (Allen et al. 1988) usually is the basis of the estimated values, but numerous variants exist depending on the weather data that are available as input (Allen 1986). The basic equations for PET describe the moisture lost from a crop canopy that completely covers the ground. Numerous assumptions are used to account for evaporation from the soil surface or a mulch layer, the portion of the ground covered by the crop, and the aerodynamic characteristics of the crop. The potential rate for evaporation from the soil surface must further be adjusted for the relative wetness of the surface and how freely moisture moves upward from lower in the soil. The storage component in the soil is positive when soil moisture increases and negative if moisture is lost. The calculations of the soil water balance are complex, in part because of the need to consider soil properties that vary with depth and management (Ritchie 1998), and the different assumptions used underlie important differences among models.

On the plant side, the potential transpiration (PET − E) establishes the upper limit for water uptake by the crop. Actual transpiration is less than the potential if insufficient soil water is readily available. The available water depends both on the amount of moisture available to the plant at different soil depths and the distribution of roots in the soil. Models typically consider a field soil described with discrete horizontal layers, which may vary in water holding capacity, moisture content, and root content, expressed in terms of mass and length. The processes of estimating available water and actual root uptake of water are again too complex for this review but are described by Ritchie (1998). Excess water in the profile or standing water may result in stress since anaerobic conditions disrupt root function. Elevated [CO_2] also reduces transpiration due to the increase in stomatal resistance with [CO_2]. In the CSM model, two response curves are used depending on whether the species is C_3 or C_4 (Fig. 4.7).

4.2.4.3 Nitrogen and Other Nutrients

Similar balance approaches are used for nitrogen and other nutrients, with modifications for conversion of the nutrient to a form in the soil solution that the roots may take up. In the case of nitrogen, this requires simulating mineralization and immobilization of nitrogen, which in turn, requires tracking levels of ammonium, nitrate and soil organic matter (Godwin and Singh 1998). For grain legumes, biological fixation of

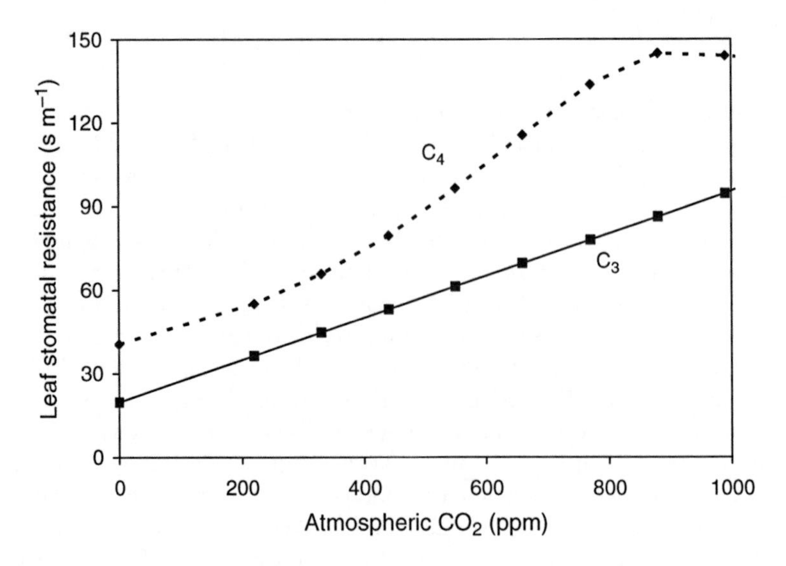

Fig. 4.7 Response to atmospheric $[CO_2]$ assumed in the CSM-CERES and CSM-CROPGRP models for leaf stomatal resistance (which affects net transpiration) for C_3 crops (e.g., common bean, peanut, soybean, barley, oats, rice and wheat) and C_4 crops (e.g., maize, millet and sorghum)

nitrogen is also considered. These soil processes respond to soil temperature and moisture status.

4.2.5 Crop Management

Besides planting per se, most crop management can be simulated by modifying the levels of certain environmental factors, especially those associated with inputs. Planting is a special case because the selection of the cultivar, planting date, and spatial arrangement (e.g., density and row spacing) set the stage for simulating the crop in the environment. Cultivar selection is especially important and is discussed separately below.

The planting date provides a starting date for simulating plant processes, although simulations often start before planting in order to track effects of pre-planting irrigations or fertilizer applications or to estimate the soil conditions prior to planting. The population and planting arrangement are important because they influence early crop development. Furthermore, some models simulate the architecture of the canopy to estimate how different portions of the canopy intercept solar radiation and how much of the soil is covered by the crop (Boote and Pickering 1994). These calculations require information on the row spacing and orientation.

In many regions, the water available through precipitation either is insufficient to support crop growth or the temporal distribution of precipitation exposes crops to periods of water deficits. Thus, where feasible, producers provide supplemental water through irrigation. In models, irrigations have an effect that is similar to rainfall. An irrigation efficiency factor may be used to reduce the amount of water that enters the ground, and water temperature and nutrient concentration are sometimes considered. For climate change research, a major concern is not so much predicting crop responses to irrigation, but determining what are reasonable assumptions for the availability of water for irrigation (Chapter 10).

Fertilizer applications are characterized by the time of application, the amount of nutrient that is provided and the type of fertilizer, and the method and depth of incorporation. Detailed models consider the chemical form of the supplied nutrient in order to characterize how readily the nutrient becomes available to plants or is lost through leaching or volatilization.

Few models directly simulate the effects of tillage. Tillage effects are sometimes modeled by altering initial conditions such as soil bulk density and a portion of plant residue on the soil surface. Full simulation typically involves changing soil physical properties such as bulk density and mixing of different soil layers (e.g., Andales et al. 2000). Models also vary in their ability to simulate crop rotations or sequences of crops over time, which requires ensuring that the soil conditions at the end of a cropping season can be used for a subsequent fallow period or the next crop in the sequence or rotation.

4.2.6 Cultivar Characteristics

Cultivar characteristics usually are embodied in a set of parameters, sometimes termed "genetic coefficients", that are thought characteristic of the species. These can characterize differences in phenology through parameters for phase durations, response to photoperiod, and, if appropriate, vernalization requirements. Parameters may also be defined for seed growth characteristics, seed composition, relative leaf size, or other traits.

Cultivar parameters are estimated by adjusting the parameters iteratively to obtain good agreement between observed and simulated values from a calibration dataset. This process requires access to extensive sets of field data, and lack of such data often constrains calibration accuracy (Anothai et al. 2008a). Recent work has shown the potential for obtaining cultivar parameters with limited data sets that are collected in state-wide variety trials (Mavromatis et al. 2001, 2002; Guerra et al. 2008) as well as in plant breeding trials (Anothai et al. 2008b). Attempts to estimate parameters from genetic information also show promise (White and Hoogenboom 1996; Messina et al. 2006; White et al. 2008). Great uncertainty remains, however, on the potential for further genetic adaptation to climate change (Ainsworth et al. 2008) and how to represent the adaptations in models. Obvious targets would involve increased heat tolerance and responses of various processes to [CO_2].

4.3 Applications in Climate Change Research

4.3.1 Basic Estimation of Climate Change Impacts

Prior to any climate change impact study, it is important that a model has been thoroughly evaluated for the crop or production system of interest. Once the evaluation is acceptable, simulating a future climate change scenario is no more difficult than simulating a crop under current or historic conditions. One needs an accurate estimate of the future weather conditions and data for the additional crop, soil and management inputs. The model is simply run with the future weather data and if considered relevant, the projected [CO_2] level. This process is the core of any simulation analysis of climate change impacts in agriculture and is found in the earliest applications of ecophysiological models (Rosenzweig 1985). However, in the selection of the appropriate model, it is important to consider whether a model will not only respond to the expected changes in climate, but also to the projected ranges. As an example, the first crop simulation models that were used for climate change studies did not include a response to the projected [CO_2] level. Most current models also have difficulties with proper responses to extreme high temperatures.

A first consideration usually is what climate change and [CO_2] scenarios to consider (see Chapters 3, 10). Comparing simulations from a single historic year with a single future year would be misleading since the results would be highly dependent on the weather conditions for that pair of years. Thus, simulations usually are conducted for sets of historic and future data that are run for 20 or more years in order to provide a more robust estimate of impacts and to account for the annual variability. Furthermore, this allows one to analyze production risk and to examine which weather variables appear to dominate predicted changes in production, resource use and environmental impact.

4.3.2 Adaptation

Production practices undoubtedly will evolve in response to climate change. However, they will also evolve with technological developments, environmental regulation, market conditions and other factors. Thus, while there is value in considering how possible adaptations in crop management might affect the impact of climate change, one must keep in mind that climate change is only one process among many that will affect future agricultural systems.

Simulating the simplest adaptations mainly involves changing planting dates, fertilizer applications, cultivars, and where applicable, irrigation practices. Effects of tillage and residue management are seldom considered, presumably because few models simulate tillage effects or their expected effects would not vary with climate change. Zhang (2005) used the Water Erosion Prediction Project (WEPP) model to compare no-till, conservation tillage and conventional tillage systems for wheat at a single site in Oklahoma and found that all three treatments increased yields about 14% under climate change. Different crop rotations or alternate crops also are seldom compared. Ideally, the options considered should be within the range that producers likely would consider adopting.

An especially difficult question is how best to assess the potential for adaptation of cultivars. Simple approaches test a limited set of existing cultivars that differ in maturity. This allows testing for response to growing season but ignores potential for changing partitioning or other growth characteristics. Gene-based approaches offer the option of examining all possible genetic combinations affecting cultivar performance (e.g., White and Hoogenboom 2005).

4.3.3 Pending Issues in Applications of Ecophysiological Models

4.3.3.1 Model Selection and Accuracy

One can imagine a well-structured process whereby a suite of potential models to be used in climate change research are tested for accuracy, considering the target crop(s) and production region. The best model or subset of models would then be used to estimate impacts. In practice, this process is seldom fully executed. The Fourth Assessment Report of the IPCC (Easterling et al. 2007) noted that previous "calls by the Third Assessment Report (TAR) to enhance crop model inter-comparison studies have remained unheeded; in fact, such activity has been performed with much less frequency after the TAR than before."

One obstacle to model intercomparisons is the lack of standardization for inputs and user interfaces. Using any given model requires a substantial effort to learn how to use the model and especially, prepare the input data for local model evaluation and scenario analysis. Thus, groups conducting impact studies often appear to select a model based on previous familiarity or general reputation of the model.

A related obstacle is that assessing model performance is less straightforward than it first seems. Comparisons of model predictions with observed data or outputs from other models require care. First, one should consider whether the evaluation datasets represent a valid sample for the target production situations. Ideally, the evaluation datasets should include conditions that test responses to elevated temperature and [CO_2], but since datasets for such conditions are seldom available, most models are evaluated primarily for current conditions. Of course, sufficient calibration data must be available to ensure that one model is not benefited over another by having been pre-calibrated to the region or production system. This can involve subtle bias including weather and soil data and assumptions concerning initial conditions and crop management. Efforts to promote standards and data sharing have met limited success (Hunt et al. 2001; Bostick et al. 2004).

Faced with apparent data limitations for evaluation, users have several options. The foremost is to seek alternative sources of data. Reports from plant breeding trials can provide a wealth of usable data (e.g., Mavromatis et al. 20002; Anothai et al. 2008b; White et al. 2008). Agronomists familiar with a given region often can provide information on expected yields and responses to inputs. Sensitivity analysis, where model inputs or parameters are varied in a systematic fashion, can also provide useful information on model performance even in the absence of comparable field observations (White et al. 2005). Examples of papers describing relatively complete evaluations are White et al. (1995), Asseng et al. (1998), Soler et al. (2007, 2008). Hartkamp et al. (2002) illustrates strategies for evaluating a model for a crop where reliable field data were especially scarce.

The statistical assumptions used to evaluate models are often suspect as well. Comparisons of observed values vs. simulated data are often analyzed with linear regression, which assumes independence of values. Any dataset from multiple locations or years or involving samples over time is likely to violate this assumption. Multiple regression can overcome some of these problems and can be used to test explicitly whether one model provides better predictions than another (White et al. 2007).

4.3.3.2 Plant Processes

Aspects of the physiology represented in models remain problematic as evidenced by the debates over responses to elevated [CO_2] (Long et al. 2006; Tubiello et al. 2007; Ziska and Bunce 2007). In simulating photosynthesis, there also is controversy concerning the role of rubisco enzyme activation in responses to heat and elevated [CO_2] (Crafts-Brandner and Salvucci 2004). Furthermore, various responses to [CO_2] seem independent of effects via photosynthesis and are not considered in models. Thomas and Harvey (1983) found that soybean leaves from plants grown at elevated [CO_2] formed an extra layer of mesophyll tissue, which would affect leaf area expansion and gas exchange properties of the leaf. Bunce (2005) found that plants exposed to continuous [CO_2] differed from plants only exposed to elevated [CO_2] during the daytime, which again is suggestive of non-photosynthetic effects of [CO_2]. Crops also vary in how elevated [CO_2] affects time to first flower (Reekie et al. 1994), but the mechanisms of such responses are unclear. Plant biology offers great promise as a source of information on the genetic control and physiology of such processes (Hammer et al. 2006). Hoogenboom and White (2003) used information on the *Tip* locus in common bean to guide improvements in simulation of the temperature and photoperiod responses.

Surprisingly few models used in climate change research quantify the crop energy balance, which requires tracing flows and transformations of energy in the soil, plant, and atmosphere. While criticizable as introducing excessive complexity, an energy balance may make simulations more robust because it provides more realistic plant and soil temperatures and ensures that energy transfers through evaporation and transpiration are realistically constrained. The *ecosys* model has successfully reproduced performance of a wheat crop grown under free-air CO_2 enrichment (FACE) conditions (Grant et al. 2001).

4.3.3.3 Model Design

Ecophysiological models evolve by having features added as understanding improves or limitations are identified. However, the modifications often emphasize expediency over robust software design. Reynolds and Acock (1997) outlined a modular modeling approach that allows components to be interchanged without requiring modifications to other parts of the model software. Modularity would greatly facilitate testing alternative physiological hypotheses related to issues such as temperature effects on photosynthesis. The approach, however, has seen only partial implementation. Individual models have become more modular in structure (e.g., Jones et al. 2003), but interchangeability of modules among models has not been attained.

The question of whether greater complexity improves model accuracy is frequently raised (e.g., Reynolds and Acock 1985; Passioura 1996). A simple model may have limited predictive capability because it does not describe a wide enough range of responses. A complex model may be inaccurate because of incorrect assumptions, programming errors, or propagation of errors from poorly estimated parameters. In the absence of rigorous model comparisons, however, it is difficult to endorse a specific level of complexity.

4.3.3.4 Application Scenarios

Issues such as what are the most accurate estimates for greenhouse gas levels or how to downscale climate change projections are dealt with in other chapters. However, many other aspects of scenario design merit review. Few studies consider how soil variability within a location might affect projections. Impact studies mainly consider crop species in isolation, yet in temperate regions the most dramatic changes in farming in temperate regions may involve changes from single crops to dual cropping and from short season cereal and oilseed crops with a spring habit to winter types.

Analyses of potential impacts of scenarios could be enhanced by greater consideration of associated responses rather than focusing on economic yield. Probably the most pressing topic is how water use might change. Assuming no adaptation in terms of cultivar type or planting date, the simplest expectation is that water use will decline due to the well documented reduction in stomatal conductance with increased [CO_2]. However, if crops are selected for greater response of net photosynthesis to [CO_2], the water-conserving response of stomata may decline, thus increasing water use. Furthermore, warmer temperatures would increase PET as well as lengthen the growing season, further increasing water use quantified on a seasonal or annual basis. Such interactions are readily simulated, but they involve plant and system responses that are still poorly understood.

4.4 Conclusion

Ecophysiological models are widely used to simulate potential impacts of climate change on agricultural systems because they reflect the best-available information on how plants respond to environmental factors and crop management. Nonetheless, use of the models involves numerous assumptions whose net effects are difficult to quantify. Results from ecophysiological models are also sensitive to the quality of inputs for cultivar traits, soil conditions, weather, and management.

Researchers concerned with impacts of climate change on agriculture should strive to understand the compromises inherent in developing and applying ecophysiological models in climate change research. Candidate models should be evaluated thoroughly and close attention paid to the accuracy of the inputs. When possible, output from more than one model should be compared and simulations should be compared with alternate approaches, such as the statistical methods discussed in the following chapters.

References

Adejuwon JO (2006) Food crop production in Nigeria. II. Potential effects of climate change. Climate Res 32:229–245

Ainsworth EA et al. (2008) Next generation of elevated [CO_2] experiments with crops: a critical investment for feeding the future world. Plant Cell Environ 31:1317–1324

Alexandrov VA, Hoogenboom G (2000) Vulnerability and adaptation assessments of agricultural crops under climate change in the Southeastern USA. Theor Appl Climatol 67:45–63

Alexandrov V, Eitzinger J, Cajic V, Oberforster M (2002) Potential impact of climate change on selected agricultural crops in north-eastern Austria. Glob Change Biol 8:372–389

Allen RG (1986) A Penman for all seasons. J Irr Drain Eng 112:348–368

Allen RG, Pereira LS, Raes D, Smith M (1988) Crop evapotranspiration – guidelines for computing crop water requirements. FAO Irrigation and drainage paper 56. Rome, Italy

Andales AA, Batchelor WD, Anderson CE (2000) Incorporating tillage effects into a soybean model. Agric Syst 66:69–98

Anothai J, Patanothai A, Pannangpetch K, Jogloy S, Boote KJ, Hoogenboom G (2008a) Reduction in data collection for determination of cultivar coefficients for breeding applications. Agric Syst 96:195–206

Anothai J, Patanothai A, Jogloy S, Pannangpetch K, Boote KJ, Hoogenboom G (2008b) A sequential approach for determining the cultivar coefficients of peanut lines using end-of-season data of crop performance trials. Field Crop Res 108:169–178

Anwar MR, O'Leary G, McNeil D, Hossain H, Nelson R (2007) Climate change impact on rainfed wheat in south-eastern Australia. Field Crop Res 104:139–147

Asseng S, Keating BA, Fillery IRP, Gregory PJ, Bowden JW, Turner NC, Palta JA, Abrecht DG (1998) Performance of the APSIM-wheat model in Western Australia. Field Crops Res 57:163–179

Bindraban PS (1999) Impact of canopy nitrogen profile in wheat on growth. Field Crop Res 63:63–77

Boote KJ, Pickering NB (1994) Modeling photosynthesis of row crop canopies. Hortscience 29:1423–1434

Boote KJ, Minguez MI, Sau F (2002) Adapting the CROPGRO legume model to simulate growth of faba bean. Agron J 94:743–756

Bostick WM, Koo J, Walen VK, Jones JW, Hoogenboom G (2004) A web-based data exchange system for crop model applications. Agron J 96:853–856

Brassard J-P, Singh B (2008) Impacts of climate change and CO_2 increase on agricultural production and adaptation options for southern Quebec, Canada. Mitigation and adaptation strategies for global change. Climate Res 13:241–265

Bunce J (2005) Seed yield of soybeans with daytime or continuous elevation of carbon dioxide under field conditions. Photosynthetica 43:435–438

Crafts-Brandner SJ, Salvucci ME (2004) Analyzing the impact of high temperature and CO_2 on net photosynthesis: biochemical mechanisms, models and genomics. Field Crop Res 90:75–85

De Wit CT (1965) Photosynthesis of leaf canopies. Agricultural research report 663. Pudoc, Wageningen

Desjardins RL, Allen LH, Lemon ER (1978) Variations of carbon dioxide, air temperature, and horizontal wind within and above a maize crop. Boundary-Layer Meteorol 14:369–380

Duncan WG, Loomis RS, Williams WA, Hanau R (1967) A model for simulating photosynthesis in plant communities. Hilgardia 38:181–205

Easterling WE, Mearns LO, Hays CJ, Marx D (2001) Comparison of agricultural impacts of climate change calculated from high and low resolution climate change scenarios: part II. Accounting for adaptation and CO_2 direct effects. Climatic Change 51:173–197

Easterling WE et al (2007) Food, fibre and forest products. In: Parry ML, Canziani OF, Palutikof JP, van der Linden PJ, Hanson CE (eds) Climate change 2007: impacts, adaptation and vulnerability. contribution of working group II to the fourth assessment report of the intergovernmental panel on climate change. Cambridge University Press, Cambridge, UK, pp 273–313

Farquhar GD, von Caemmerer S, Berry JA (1980) A biochemical model of photosynthetic CO_2 assimilation in leaves of C_3 species. Planta 149:78–90

Gitay H, Brown S, Easterling W, Jallow B (2001) Ecosystems and their goods and services. In: McCarthy JJ, Canziani OF, Leary NA, Dokken DJ, White KS (eds) Climate change 2001: impacts, adaptation, and vulnerability. Third assessment report of the intergovernmental panel on climate change. Cambridge University Press, Cambridge, UK

Godwin DC, Singh U (1998) Cereal growth, development and yield. In: Tsuji GY, Hoogenboom G, Thornton PK (eds) Understanding options for agricultural production. Kluwer, Dordrecht, the Netherlands

Grant RF, Kimball BA, Brooks TJ, Wall GW, Pinter PJ Jr, Hunsaker DJ, Adamsen FJ, Lamorte RL, Leavitt SW, Thompson TL, Matthias AD (2001) Modeling interactions among carbon dioxide, nitrogen, and climate on energy exchange of wheat in a free air carbon dioxide experiment. Agron J 93:638–649

Grimm SS, Jones JW, Boote KJ, Hesketh JD (1993) Parameter estimation for predicting flowering date of soybean cultivars. Crop Sci 33:137–144

Guerra LC, Hoogenboom G, Garcia y Garcia A, Banterng P, Beasley Jr JP (2008) Determination of cultivar coefficients for the CSM-CROPGRO-Peanut model using variety trial data. Trans ASAE 51:1471–1481

Hammer G et al (2006) Models for navigating biological complexity in breeding improved crop plants. Trends Plant Sci 11:587–593

Hanks J, Ritchie JT (1991) Modeling plant and soil systems. ASSA, CSSA, SSSA, Madison, WI

Hartkamp AD, Hoogenboom G, White JW, Gilbert R, Benson T, Barreto HJ, Gijsman A, Tarawali S, Bowen W (2002) Adaptation of the CROPGRO growth model to velvet bean as a green manure cover crop: II. Model testing and evaluation. Field Crop Res 78:27–40

Hay R, Porter J (2006) The physiology of crop yield, 2nd edn. Blackwell, Oxford, UK

Hoogenboom G, White JW (2003) Improving physiological assumptions of simulation models by using gene-based approaches. Agron J 95:82–89

Hoogenboom G, Jones JW, Wilkens PW, Porter CH, Batchelor WD, Hunt LA, Boote KJ, Singh U, Uryasev O, Bowen WT, Gijsman AJ, du Toit A, White JW, Tsuji GY (2004) Decision support system for agrotechnology transfer version 4.0 [CD-ROM]. University of Hawaii, Honolulu, HI

Hunt LA, White JW, Hoogenboom G (2001) Agronomic data: advances in documentation and protocols for exchange and use. Agric Syst 70:477–492

Jamieson PD, Brooking IR, Semenov MA, McMaster GS, White JW, Porter JR (2007) Reconciling alternative models of phenological development in winter wheat. Field Crop Res 103:36–41

Jones PG, Thornton PK (2003) The potential impacts of climate change on maize production in Africa and Latin America in 2055. Global Environ Chang 13:51–59

Jones CA, Bland WL, Ritchie JT, Williams JR (1991) Simulation of root growth. In: Hanks J, Ritchie JT (eds) Modeling plant and soil systems. ASA-CSSA-SSSA, Madison, WI, pp 91–123

Jones JW et al (2003) The DSSAT cropping system model. Eur J Agron 18:235–265

Long SP, Ainsworth EA, Leakey ADB, Nosberger J, Ort DR (2006) Food for thought: lower-than-expected crop yield stimulation with rising CO_2 concentrations. Science 312:1918–1921

Loomis RS, Rabbinge R, Ng E (1979) Explanatory models in crop physiology. Annu Rev Plant Physiol 30:339–367

Mavromatis T, Boote KJ, Jones JW, Irmak A, Shinde D, Hoogenboom G (2001) Developing genetic coefficients for crop simulation models with data from crop performance trials. Crop Sci 41:40–51

Mavromatis T, Boote KJ, Jones JW, Wilkerson GG, Hoogenboom G (2002) Repeatability of model genetic coefficients derived from soybean performance trails across different states. Crop Sci 42:76–89

Messina CD, Jones JW, Boote KJ, Vallejos CE (2006) A gene-based model to simulate soybean development and yield responses to environment. Crop Sci 46:456–466

Minguez MI, Ruiz Ramos M, Diaz Ambrona CH, Quemada M, Sau F (2007) First-order impacts on winter and summer crops assessed with various high-resolution climate models in the Iberian Peninsula. Climatic Change 81:343–355

Monteith JL, Unsworth MH (1990) Principles of environmental physics. Edward Arnold, London

Passioura JB (1996) Simulation models: science, snake oil, education, or engineering? Agron J 88:690–694

Penning De Vries FWT, Brunsting AHM, Van Laar HH (1974) Products, requirements and efficiency of biosynthesis a quantitative approach. J Theor Biol 45:339–377

Porter JR, Gawith M (1999) Temperatures and the growth and development of wheat: a review. Eur J Agron 10:23–36

Reekie JYC, Hickleton PR, Reekie EG (1994) Effects of elevated CO_2 on time to flowering in four short-day and four long-day species. Can J Bot 72:533–538

Reynolds JF, Acock B (1985) Predicting the response of plants to increasing carbon dioxide: a critique of plant growth models. Ecol Model 29:107–129

Reynolds JF, Acock B (1997) Modularity and genericness in plant and ecosystem models: modularity in plant models. Ecol Model 94:7–16

Ritchie JT (1998) Soil water balance and plant water stress. In: Tsuji GY, Hoogenboom G, Thornton PK (eds) Understanding options for agricultural production. Kluwer, Dordrecht, the Netherlands, pp 41–54

Ritchie JT, NeSmith DS (1991) Temperature and crop development. In: Hanks J, Ritchie JT (eds) Modeling plant and soil systems. ASSA, CSSA, SSSA, Madison, WI, pp 5–30

Rosenzweig C (1985) Potential CO_2-induced climate effects on North American wheat-producing regions. Climatic Change 7:367–389

Singh U, Matthews RB, Griffin TS, Ritchie JT, Hunt LA, Goenaga JT (1998) Modeling growth and development of root and tuber crops. In: Tsuji GY, Hoogenboom G, Thornton PK (eds) Understanding options for agricultural production. Kluwer, Dordrecht, the Netherlands, pp 1

Soler CMT, Sentelhas PC, Hoogenboom G (2007) Application of the CSM-CERES-Maize model for planting date evaluation and yield forecasting for maize grown off-season in a subtropical environment. Eur J Agron 27:165–177

Soler CMT, Maman N, Zhang X, Mason SC, Hoogenboom G (2008) Determining optimum planting dates for pearl millet for two contrasting environments using a modelling approach. J Agric Sci 146:445–459

Spaeth SC, Sinclair TR (1985) Linear increase in soybean harvest index during seed-filling. Agron J 77:207–211

Stockle CO, Donatelli M, Nelson R (2003) CropSyst, a cropping systems simulation model. Eur J Agron 18:289–307

Sung S, Amasino RM (2004) Vernalization and epigenetics: how plants remember winter. Curr Opin Plant Biol 7:4–10

Thomas JF, Harvey CN (1983) Leaf anatomy of four species grown under long-term continuous CO_2 enrichment. Bot Gaz 144:303–309

Tingem MR, Rivington M, Bellocchi G, Azam-Ali S, Colls J (2008) Effects of climate change on crop production in Cameroon. Climate Res 36:65–77

Tsuji GY, Hoogenboom G, Thornton PK (eds.) (1998) Understanding options for agricultural production, Kluwer, Dordrecht, the Netherlands

Tubiello FN, Amthor JS, Boote KJ, Donatelli M, Easterling W, Fischer G, Gifford RM, Howden M, Reilly J, Rosenzweig C (2007) Crop response to elevated CO_2 and world food supply: a comment on "Food for Thought..." by Long et al., Science 312, 1918–1921, 2006. Eur J Agron 26:215–223

von Caemmerer (2000) Biochemical models of leaf photosynthesis. CSIRO, Collingwood, Australia

White JW, Hoogenboom G (1996) Simulating effects of genes for physiological traits in a process-oriented crop model. Agron J 88:416–422

White JW, Hoogenboom G (2005) Integrated viewing and analysis of phenotypic, genotypic, and environmental data with "GenPhEn Arrays". Eur J Agron 23:170–182

White JW, Hoogenboom G, Jones JW, Boote KJ (1995) Evaluation of the dry bean model Beangro V1.01 for crop production research in a tropical environment. Exp Agric 31:241–254

White JW, Hoogenboom G, Hunt LA (2005) A structured procedure for assessing how crop models respond to temperature. Agron J 97:426–439

White JW, Boote KJ, Hoogenboom G, Jones PG (2007) Regression-based evaluation of ecophysiological models. Agron J 99:419–427

White JW, Herndl M, Hunt LA, Payne TS, Hoogenboom G (2008) Simulation-based analysis of effects of *Vrn* and *Ppd* loci on flowering in wheat. Crop Sci 48:678–687

Williams JR, Jones CA, Kiniry JR, Spanel DA (1989) The EPIC crop growth model. Trans ASAE 32:497–511

Zhang X-C (2005) Spatial downscaling of global climate model output for site-specific assessment of crop production and soil erosion. Agricultural and Forest Meteorology 135:215–229

Ziska LH, Bunce JA (2007) Predicting the impact of changing CO_2 on crop yields: some thoughts on food. New Phytol 175:607–618

Chapter 5
Crop Responses to Climate: Time-Series Models

David Lobell

Abstract Time series of annual crop production levels, at scales ranging from experimental trials to regional production totals, are widely available and represent a useful opportunity to understand crop responses to weather variations. This chapter discusses the main techniques of building models from time series and the tradeoffs involved in the many decisions required in the process. A worked example using United States maize production is used to illustrate key concepts.

5.1 Introduction

The task of predicting crop responses to climate would be easy if crop yield were determined by a single and simple biological process. The reality, of course, is more complex. Crop growth and reproduction are governed by many interacting processes that present an enormous challenge to efforts at prediction. Many of the most relevant processes have been outlined in Chapter 4, which describes efforts to develop models that capture the essence of each process without being too complex to prevent reliable model calibration and applications.

An alternative to this process-based approach is to rely on the statistical relationships that emerge between historical records of crop production and weather variations. In short, we observe the past and use it to build models to inform the future. From the outset, it should be clear that purely statistical approaches, whether based on time series as discussed in this chapter or cross-sectional data as discussed in the next, are not inherently better or worse than more process-based approaches. There are some disadvantages, such as difficulty in extrapolating beyond historical extremes, as well as some advantages, such as limited data requirements and the potential to capture effects of processes that are relatively poorly understood, such as pest dynamics.

It should also be clear that statistical approaches cannot proceed successfully without some consideration of the underlying processes. For example, the choice of which months of weather to consider will depend on the growing season of the

D. Lobell
Stanford University, CA, USA

D. Lobell and M. Burke (eds.), *Climate Change and Food Security*,
Advances in Global Change Research 37, DOI 10.1007/978-90-481-2953-9_5,

crop, and the choice of what climate variables to use will depend on the processes thought to be most important. Such considerations will be explained in more detail below. The general point, and one that is often confused in the existing literature, is that the distinction between "process-based" and "statistical" models is somewhat arbitrary. All process-based models have some level of empiricism, and all statistical models have some underlying assumptions about processes.

This chapter seeks to describe time series based approaches to crop modeling, highlighting the important decisions that can affect the outcomes. Time series models have been widely used to evaluate the impacts of climate variability and change on crop production. They are particularly useful in situations where there is insufficient data to calibrate more process-based models, and where detailed spatial datasets are not available, both of which are accurate descriptions of the situation in many developing countries. Their main requirement is the availability of sufficiently long time series (at least 20 years) of both weather and crop harvests.

5.2 A Worked Example: U.S. Maize Yields

To help guide the discussion and illustrate the time series modeling process, we will use a dataset for US maize yields from 1950–2005. The United States Department of Agriculture (USDA) has recorded average yields for each county since early in the twentieth century, and in many cases since the late 1800s. In addition, the United States has arguably the most complete weather records of any country for the twentieth century. Here we have averaged yields over all counties east of the 100° W meridian, which separates mostly irrigated maize in the West from mostly rainfed maize in the East. The average was weighted by the area sown to maize in each county, so that yields in counties with high acreage were proportionally more important. Weather data from individual stations were similarly weighted.

5.3 Common Issues in Time Series Modeling

5.3.1 Spatial and Temporal Extent

The choice of restricting our time series to east of 100° W highlights one of the first decisions in time series modeling – the spatial extent over which yields and weather are averaged. This scale can range from individual fields (if the data are available) to entire regions. As scales become bigger, datasets are often more reliable and available. However, aggregating areas too large will result in combining fields that actually behave quite differently. Consider, for example, two adjacent areas, one of which prefers cooler climates and the other of which prefers warmer climates. If these two are combined, then the average yields could show no effect of climate variations even though there is a true response in each region. If the only goal is to understand yield responses at the broader scale, then one might be satisfied with this result. But, in general, we prefer when possible to work with regions that are relatively homogeneous in nature. Separating irrigated from rainfed crops is almost always a good idea, given the very different nature of response to rainfall.

The scale issue also extends to the temporal dimension. Just as maize in California may be functionally a very different crop than maize in Georgia, maize in 1950 was potentially very different in its climate response than maize in 2000. For example, Zhang et al. (2008) demonstrate that the correlation between rice yields and temperature in China switched from being negative before 1980 to positive after 1980. The explanation in this instance was that irrigation was much more widespread after 1980, allowing crops to take advantage of the drier, sunnier conditions that led to water stress for rainfed crops. The time period for time series analysis should therefore be restricted to periods over which management was reasonably constant, particularly management factors such as irrigation that can strongly influence climate responses. One approach to ensure a stationary relationship between climate and yields is to perform the analysis for the first and second half of the record, and then compare the results.

5.3.2 Trend Removal

Figure 5.1a shows average yields over the study period. The most obvious feature of this time series, and time series of yields for most crops in most regions, is the highly significant positive trend with time. This trend results largely from improvements in technology, such as adoption of modern hybrid cultivars and increased use of fertilizer. Given that so much of yield variation between years in different parts of the record occurs because of technology differences, the effect of climate is difficult to discern from the raw yield data. For that reason, one nearly always performs a de-trending of the data to remove the influence of technology. There are several ways to do this, none of them clearly optimal. The first is to approximate the trend in technology with a polynomial fit, and take the yield anomalies from this trend. For most crops the technology trend can be approximated with a first order polynomial (linear trend).

Figure 5.1b illustrates the yield anomalies from a linear trend for the maize time series. The anomalies are much larger in absolute value for the latter part of the record, a common occurrence in yield time series. This change in variance from the beginning to end of the record, known as heteroskedasticity, violates some of the basic assumptions of many statistical techniques such as linear regression. To correct for this, yields are often expressed on a log basis, which means that anomalies represent percent differences from the trend line rather than absolute differences, since log (a) − log (b) = log (a/b). As shown in Fig. 5.1c, use of log yields rather than absolute yields removes most of the problem with heteroskedasticity.[1]

---

[1] However, note that if the yield anomalies in Fig. 5.1b showed no sign of heteroskedasticity, then introducing the log transformation could lead to heteroskedasticity by suppressing values at the beginning of the record.

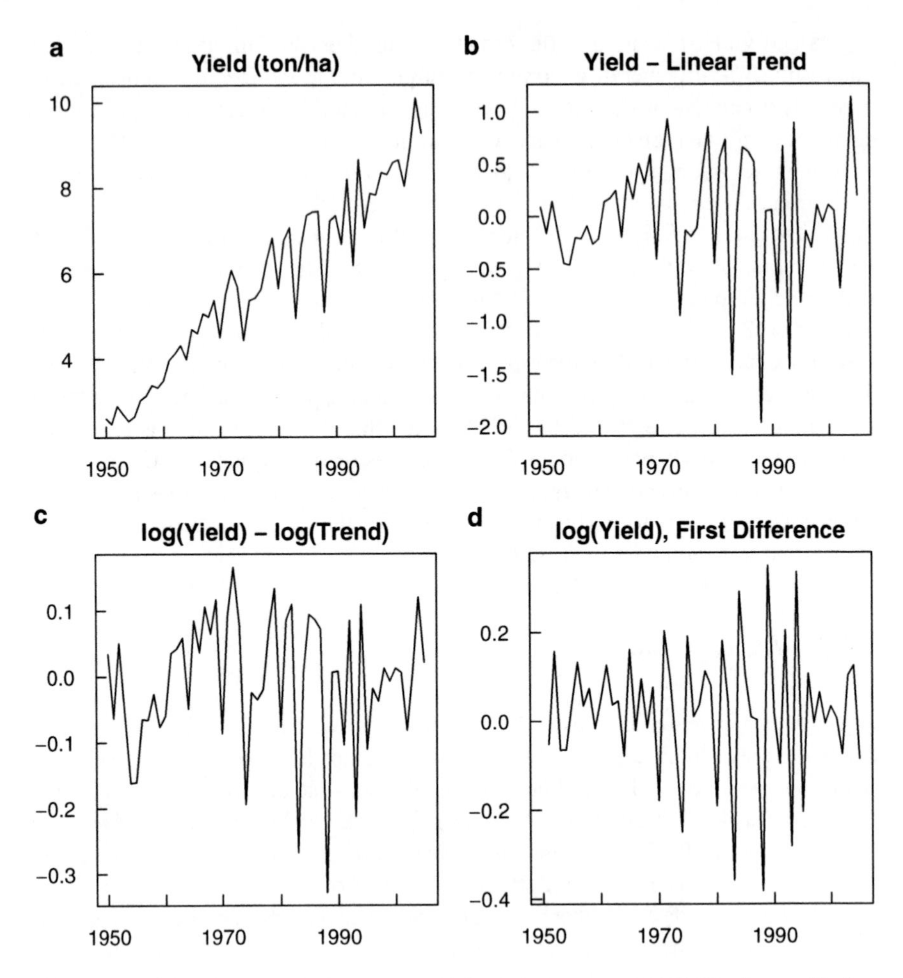

Fig. 5.1 (**a**) Time series of maize yields in US for counties east of 100° W, shown with three common methods of detrending (**b–d**)

It is frequently the case that yield trends are obviously not linear, as demonstrated for two cases in Fig. 5.2. In this situation, fitting a linear trend may cause serious errors, and one can resort instead to higher order polynomials. A more flexible approach, and one that is commonly used in time series analysis, is to transform the data to first-differences as shown in Fig. 5.1d, where from each value one subtracts the value in the previous year. In this case, the subsequent analysis focuses only on year-to-year changes so that effects of long-term trends are minimized. Any predictor variables must then also be transformed to first-differences in order to compare with yields.

A final approach to account for technology is not to remove a trend, but rather to include a term for year (and possibly year-squared) in subsequent regression analysis. One could also include explicit technology proxies, such as fertilizer rate or percent of growers using modern cultivars.

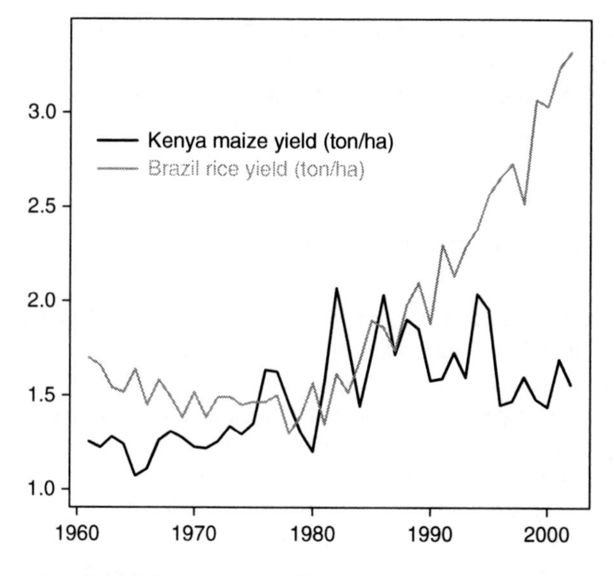

Fig. 5.2 Time series of yields in two cases with a nonlinear trend

Table 5.1 Summary of regression model results for different methods of detrending

Response variable	Predictor variable(s)	Model R^2	Yield sensitivity (mean ± 1 s.d.% °C^{-1})
Yield	Avg. temperature and year	0.92	−3.8 ± 2.0
Log (Yield)	Avg. temperature and year	0.90	−4.5 ± 2.5
Yield–Trend	Avg. temperature	0.06	−3.7 ± 1.9
Log (Yield)–Log (Trend)	Avg. temperature	0.10	−4.4 ± 1.9
Yield, first difference	Avg. temperature, first difference	0.16	−7.6 ± 2.4
Log (Yield), first difference	Avg. temperature, first difference	0.16	−6.8 ± 2.3

In summary, for any yield time series of considerable length, accounting for technology trends is essential, and many approaches exist toward this end. How important is this decision in the final analysis? Table 5.1 summarizes the results of simple linear regressions with average growing season (April–September) temperature as the predictor variable and various representations of yield as the response variable. The model R^2 indicates that regressions using first differences tend to have higher explanatory power than those based on anomalies. Models that use raw yields and include a time trend have, of course, much higher R^2 because the effect of technology has not been previously removed but is included in the model.

The key aspect of these models is the predicted response to temperature, which is expressed as the % change in yield for a 1°C increase. The results can vary by a factor of 2, with the smallest effect found when using raw yields with a time term, and the biggest effect using first differences of raw yields. Note that the effect of using log relative to absolute yields can either increase or reduce the model R^2 and inferred yield sensitivity, while the effect of using first-differences tends to increase both in this example.

5.3.3 Climate Variable Selection

The above example used average growing season temperature, which is indeed a very common measure of growing season weather. However, there are many other defensible variables to use in place of or in addition to this value. We distinguish here between two main choices: variable type and temporal scale. Variable type decisions involve, for instance, whether to include a term related to temperature, one for precipitation, and/or one for solar radiation or some other meteorological variable. Temporal scale decisions include extent (i.e., what length of growing season to consider) and resolution (i.e., how many intervals within the growing season to include). For example, while we defined the growing season as April–September, one could argue that March–August or June–September is a better definition. For resolution, many have argued that intra-seasonal variations in weather can be as important as averages (e.g., Thompson 1986; Hu and Buyanovsky 2003; Porter and Semenov 2005). Heat or rainfall during critical flowering stages for example, may be as or more important than average conditions. Again, while this is certainly true to some extent, the key question is how much the final analysis is affected by this decision.

One aspect of intra-seasonal variation is the length of time the crop spends above critical heat thresholds. For maize, it is commonly thought that temperatures above 30°C are particularly bad for crop development and growth (see Chapter 4). With hourly data, one can compute the number of hours spent above some threshold for the entire growing season in addition or in lieu of using growing season averages. Such decisions depend a great deal on the availability of fine scale meteorological measurements. In many parts of the world, reliable data are only available for monthly averages (briefly discuss here the approach to deriving degree days).

Figure 5.3 illustrate three climate variables for US maize: average growing season temperature and precipitation, and degree days above 30°C (GDD30), all plotted against each other and yield anomalies. The numbers below the diagonal in Fig. 5.3 indicate the correlation coefficient between the pair of variables. In this example, average temperature shows a significant correlation with yields, but less so than GDD30. Precipitation exhibits a slight positive correlation with yields and negative correlations with both temperature measures.

An important point illustrated in Fig. 5.3 is that different climate variables are often highly correlated with each other, such as average temperature and GDD30 in this example. Thus it is impossible to say exactly how much of the observed correlation between yields and average temperatures is due to a real effect of average conditions, and how much is due to a real effect of very hot days or reduced precipitation that happens to be correlated with average temperatures.

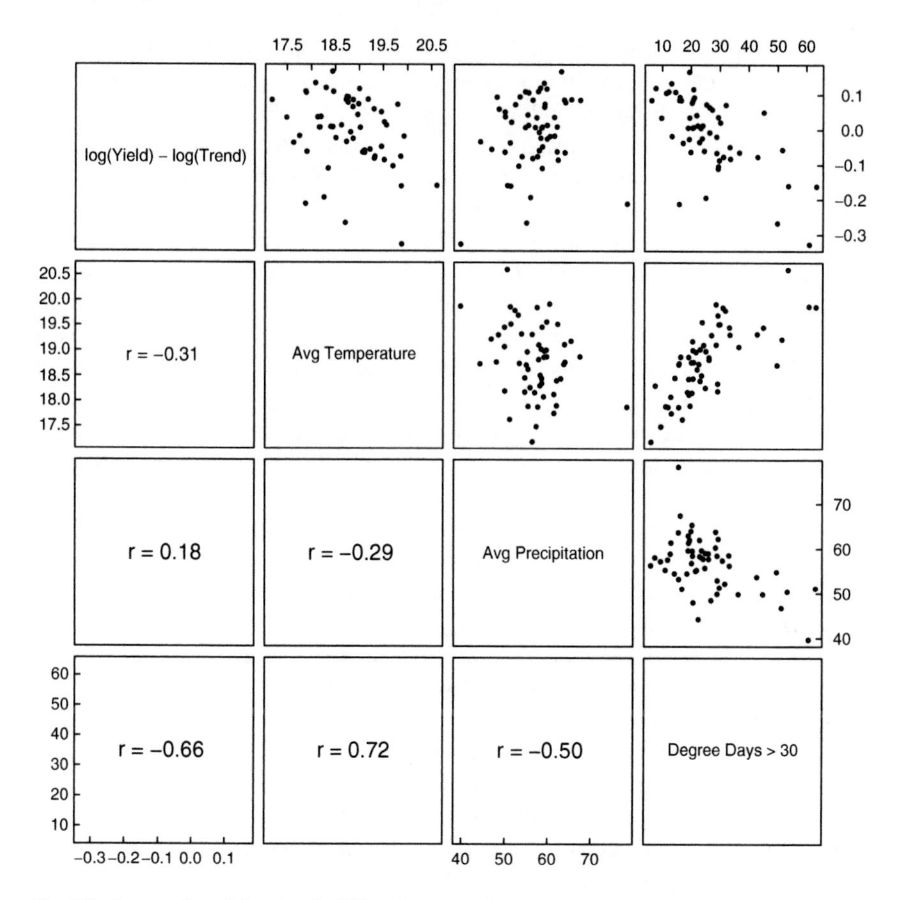

Fig. 5.3 Scatterplot of data for the US maize example

This problem of colinearity is common in statistical analysis, and often makes it impossible to attribute yield changes to a single climate variable. The obvious risk is that one may attribute yield losses to one variable when in fact another variable is the true culprit. The best approach to minimizing colinearity is to obtain samples where the climate variables are not highly correlated. For example, although growing season daytime and nighttime average temperatures are often very highly correlated, there are some locations in the world where this is not the case. Lobell and Ortiz-Monasterio (2007) focused on three such regions to evaluate the response of wheat yields to night and day temperatures.

A useful method for gauging the effect of colinearity is to evaluate partial correlation coefficients, i.e., the correlation between yield and a climate variable after the correlations with all other variables have been removed. Similarly, one can compute regressions between a variable and the residuals from a regression of yield on all other variables. Comparison of this value with the coefficient from an ordinary multiple regression will provide some measure of the role that colinearity plays.

Overall, colinearity is perhaps the biggest obstacle to time series modeling. In some cases it may be possible to distinguish between apparent and true effects on yield with knowledge of biological processes. More likely, this distinction is subjective and subject to disagreement. In an analysis of experimental rice yield responses to warming, for instance, Peng et al. (2004) reported a roughly 10% loss of yield for each degree of nighttime warming based on time series analysis. A subsequent analysis by Sheehy et al. (2006) used the rice process-based model ORYZA2000 to demonstrate that roughly half of the perceived effect of temperature could actually be due to changes in solar radiation, which are negatively correlated with nighttime temperature in this location. Similarly, Lobell and Ortiz-Monasterio (2007) compared statistical models with CERES-Wheat simulations to show that correlations of solar radiation and nighttime temperature can confound interpretation of statistical models. In the end, only controlled experiments can be used to uniquely identify the effect of a single variable when all others are held constant.

A related point illustrated by Fig. 5.3 is that omission of important variables can bias results. Maize yield correlates much more strongly with GDD30 than average growing season temperature in this region. Yet measures of exposure to extreme heat such as GDD30 have not been widely used, with most studies focused a priori on weekly or monthly averages. The choice of which variables to consider is often dictated by data availability – there are few regions in the world where reliable sub-daily data on temperatures extend back prior to 1980. There are similarly few good datasets on solar radiation, which as discussed above can be an important omitted variable because it is often correlated with temperature and rainfall.

Only by comparing results with and without the inclusion of variables such as GDD30 or solar radiation can we estimate the bias that their omission introduces in specific locations. Moreover, only by repeating these studies for a large number of locations can we make more general statements about the importance of these factors for future impacts, although strong claims for the importance of extreme events are frequently heard (Easterling et al. 2007).[2] It should also be clear that the importance of different variables may depend on the time scale for which projections are being made. For example, GDD30 may initially increase slowly as temperatures rise but more rapidly as average temperatures approach 30°C.

To summarize, time series methods are hampered by frequently high correlations between climate variables. In cases where two correlated variables are both included in the model, attribution of yield changes to any single variable is difficult if not impossible. In cases where an important variable is omitted, there is risk of attributing too much importance to a correlated variable included in the model. Even when the omitted variable is not correlated with included variables, there is a risk that its omission will miss an important effect of climate on yields.

---

[2] The recent IPCC Fourth Assessment Report states that "Projected changes in the frequency and severity of extreme climate events will have more serious consequences for food and forestry production, and food insecurity, than will changes in projected means of temperature and precipitation (high confidence)."

One may wonder at this point why we do not typically just include all possible climate variables in a regression analysis. As already stated, one common reason for omitting variables is lack of reliable data. More fundamental is the fact that increasing model complexity by adding more and more variables will eventually result in a model that is over-fit to the data, including the noise present in the data, and has worse predictive skill than a model with fewer variables. The balance between including enough but not too many variables is known in statistics as the bias-variance tradeoff, and places a premium on choosing variables wisely. As mentioned, knowledge of the biological processes that control crop growth and reproduction can be of tremendous value in the search for the "right" variables.

5.3.4 Functional Forms

Functional form refers to the type of relationship specified between a predictor variable, X, and a yield response variable, Y. The form could be a polynomial relationship, such as Eqs. (5.1) and (5.2), or an exponential relationship such as Eq. (5.3).

$$Y = \beta_0 + \beta_1 X \tag{5.1}$$

$$Y = \beta_0 + \beta_1 X + \beta_2 X^2 \tag{5.2}$$

$$\mathrm{Log}(Y) = \beta_0 + \beta_1 X \tag{5.3}$$

Several other classes of equations could also be used, such as regression trees, neural networks, or Mitscherlich equations. The most common forms used for modeling yield responses to weather are the linear model of Eq. (5.1) and the quadratic model of Eq. (5.2).

A useful way to determine the appropriate functional form is to examine a scatter plot of the data, such as in Fig. 5.3. One can also use statistical tests to determine whether a squared term significantly improves the model. A squared term can be very useful when there is an optimum temperature or precipitation amount that falls within the observed data. Thompson (1986), for instance, found that yields were reduced for departures from average June temperatures in five Corn Belt states, whether the departures were towards cooler or warmer weather.

Adding a squared term does not always help, however, as it is adds to the model complexity and can lead to overfitting and lower predictive skill. In general, we have found that higher order terms are more useful as the range of temperatures or precipitation that the crop experiences becomes wider. The reason is illustrated in Fig. 5.4: although no weather variable ever has a truly linear effect,[3] a linear approximation can be appropriate over a limited range of the weather variable. In this example, yield exhibits a nonlinear response to temperature, but this response is well approximated

[3] Extremely low and high values are nearly always bad for crops, so that the optimum value is found somewhere between.

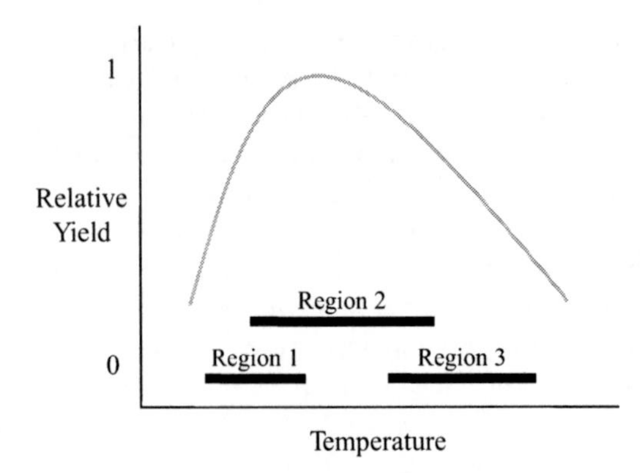

Fig. 5.4 A hypothetical relationship between temperature and yield. The range of temperatures experienced in a region will determine whether a linear approximation is appropriate

in regions 1 and 3 by a linear fit, since they are always on the cool and warm side of the optimum, respectively. In contrast, region 2 includes both temperatures where warming is strongly beneficial and temperatures where warming is quite harmful. Thus, a non-linear function would be necessary in region 2 but not the others.

It follows from the fact that the appropriateness of linear approximations depends on the range of weather experienced that the appropriateness will vary with the choice of model scale, since averages over large regions will show less variation from year to year than will averages over smaller areas. Linear models are therefore usually more appropriate when looking at national or regional time series than when looking at individual counties or states. As a case in point, the relationship between temperature and yield in Fig. 5.3 appears roughly linear even though at the state scale maize yields can exhibit strong nonlinear relationships with weather (Thompson 1986; Schlenker and Roberts 2006). Thus, as with the previous issues, the best choice for functional form will vary with the particular crop, location, and scale of interest.

5.3.5 Data Quality and Regression Bias

The example of US maize yields represents perhaps the most accurate long (50+ years) time series available on both crop yield and climate anywhere in the world. In many countries of prime interest for food security, the quality of data can be considerably worse. The crop production database of the Food and Agriculture Organization of the United Nations (FAO), for instance, contains an enormous wealth of information but much of it is visibly suspect. Reported yields are often identical for 3 or more years in a row, and areas can change dramatically in a single year. Errors in the response variable tend to inflate the standard error of coefficients in a regression model, but as long as the errors are random they should not introduce bias into the estimation procedure (Chatfield 1996). Errors in the predictor variables - in our case climate measurements – are a more serious concern because they tend to bias the coefficients towards zero. This phenomenon is known as regression bias, and though several methods exist to attempt to correct for it (Frost and Thompson 2000) its effects are often not well understood.

5.4 Projecting Impacts of Climate Change with Time Series Models

Once a model has been calibrated with time series data, it can be used to predict yield responses to any hypothetical amount of climate change. (Chapter 3 describes approaches for downscaling climate projections for input into crop models.) For example, temperature and precipitation changes from climate model simulations can be used to generate new values of the relevant predictor variables, which the regression model then translates to yields.

There are, however, three extremely important caveats to the use of time series models for simulating yield responses to climate change, even for the analyst who has successfully navigated the issues described in the previous section. The first is common to all cases of statistical model prediction, and relates to the problem of extrapolating the model beyond the range of calibration data. In particular, as the climate warms growing season temperatures may increasingly exceed the warmest year contained in the historical data used to fit the statistical model. Figure 5.5 illustrates this for the current example: as temperatures rise fewer and fewer predictions will reside in the calibration domain.

The simplest approach to avoiding extrapolation errors is to use statistical models only for the relatively near term where the vast majority of years have historical precedents. For example, if we set an arbitrary threshold that no more than 25% of

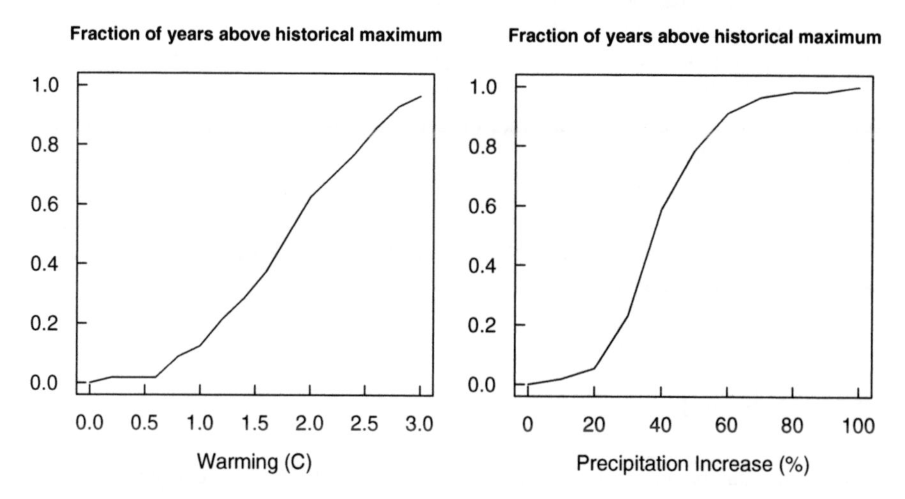

Fig. 5.5 Fraction of years that exceed historical maximum for the US maize example when temperature (*left*) or precipitation (*right*) is increased by different amounts

years should be warmer than the warmest year on record, then based on Fig. 3.5 we would only make projections out to 1.3°C. This corresponds roughly to average projections for 2030, which is still a useful period to analyze. However, using time series based models to make projections to 2080 – where climate model projections commonly exceed 3°C of warming – would be misguided, and in that case other approaches would be more appropriate.

Another approach to address extrapolation error is to implement several different methods of extrapolation to gauge whether results are sensitive to predictions made outside the calibration domain. For example, one can contrast a conservative approach of truncating yields to historical extremes, with a more aggressive approach of allowing yields to extrapolate to zero (Lobell et al. 2006). The point at which the two methods diverge provides a measure of when the time series model is on shaky ground.

The second caveat is another common one in statistics and involves the assumption of stationarity – that relationships observed in the past also apply to the future. As crop varieties and management systems change, however, the response of yields to variations in weather may also change. An example already mentioned is when irrigation is introduced into currently rainfed areas. As with extrapolation, the assumption of stationarity becomes more questionable as the time horizon of projections extends further into the future.

The final and perhaps most serious caveat is the use of models based on year-to-year variations in weather to predict responses to gradual changes in climate. An economist would refer to this as equating short-run and long-run effects, which ignores the ability of humans to adapt to system shocks. For agricultural systems, we attribute the difference between weather and climate responses to the ability of farmers to (1) perceive and (2) adapt to a changing climate. Some have gone so far as to argue that the response to climate can be opposite in sign to that for weather (Hansen 1991), while others argue that adaptation will be very difficult and not entirely effective.

A detailed discussion of adaption is presented in Part 3 of this book. The only point made here is that applying time series based models to projection of climate change implicitly assumes that no adaptation will take place. Note that this assumption does not have to be true for the projections to be useful. One goal of projections, for example, can be to identify where the biggest threats are to agriculture if we do not adapt, in order to guide short-term investments in adaptation (Lobell et al. 2008). Also, comparing time series based projections with those that incorporate adaptation can provide a useful measure of the potential impact of adaptation, a point explored further in the next chapter.

5.5 Summary

Time series can be an invaluable resource for understanding the aggregate response of crop production to variations in climatic conditions. Models based on time series depend not only on the data, but on several choices faced in the modeling process.

Most prominent among these are choosing the spatial and temporal extent of the time series data, method of detrending, types and temporal resolution of climatic variables, and the specification of the functional relationship between climate and yield. Poor choices for any of these can potentially lead to invalid estimates of climate responses.

Two general principles are especially useful for time series models. First, the analyst should always plot the data at each step, to examine features such as colinearity, heteroskedasticity, and nonlinearities, rather than rely exclusively on model summaries provided by common software packages. Second, when a choice between two alternatives is not apparent, the analysis should be tried both ways and the results compared. This is analogous to using multiple process-based models that have different but equally defensible assumptions to evaluate model uncertainty.

Users of time series models should be keenly aware that adaptation can, in principle, cause fundamentally different responses to weather and climate. As time series models are based on year-to-year variations in weather, their application to future scenarios of climate change embody an assumption of no adaptation. This can be useful in many situations, especially when results are contrasted with estimates of impacts that include adaptation, but the assumption should be kept explicit at all times.

A summary of the key points of this chapter are given below.

- Time series models can be extremely useful for projections for the next 20–30 years, when adaptation is likely to be small and climate is not too far from current conditions. Beyond that, the extrapolation of past relationships to the future becomes more tenuous.
- The most pervasive challenge in time series modeling is co-linearity between the major climate variables known to affect crops, namely temperature, precipitation, and solar radiation.
- There is no single best approach to time series modeling, as optimal decisions will depend on location, crop, and scale. Comparison of results from multiple alternative specifications can be a useful measure of uncertainty.

References

Chatfield C (1996) The analysis of time series. CRC, Boca Raton, FL

Easterling W, Aggarwal P, Batima P, Brander K, Erda L, Howden M, Kirilenko A, Morton J, Soussana JF, Schmidhuber J, Tubiello F (2007). Chapter 5: food, fibre, and forest products. in: climate change 2007: impacts, adaptation and vulnerability contribution of working group II to the fourth assessment report of the intergovernmental panel on climate change. Cambridge University Press, Cambridge, UK and New York, NY, USA

Frost C, Thompson SG (2000) Correcting for regression dilution bias: comparison of methods for a single predictor variable. J R Stat Soc A Sta 163(2):173–189

Hansen L (1991) Farmer response to changes in climate: the case of corn production. J Agr Econ Res 43(4):18–25

Hu Q, Buyanovsky G (2003) Climate effects on corn yield in Missouri. J. Appl Meteorol 42(11):1626–1635

Lobell DB, Ortiz-Monasterio JI (2007) Impacts of day versus night temperatures on spring wheat yields: a comparison of empirical and CERES model predictions in three locations. Agron J 99(2):469–477

Lobell DB, Field CB, Cahill KN, Bonfils C (2006) Impacts of future climate change on California perennial crop yields: model projections with climate and crop uncertainties. Agric For Meteorol 141(2–4):208–218

Lobell DB, Burke MB, Tebaldi C, Mastrandrea MD, Falcon WP, Naylor RL (2008) Prioritizing climate change adaptation needs for food security in 2030. Science 319(5863):607–610

Peng S, Huang J, Sheehy J, Laza R, Visperas R, Zhong X, Centeno G, Khush G, Cassman K (2004) Rice yields decline with higher night temperature from global warming. Proc Natl Acad Sci U S A 101(27):9971–9975

Porter JR, Semenov MA (2005) Crop responses to climatic variation. Philos Trans R Soc B-Biol Sci 360(1463):2021–2035

Schlenker W, Roberts MJ (2006) Nonlinear effects of weather on corn yields. Rev Agric Econ 28(3):391–398

Sheehy JE, Mitchell PL, Ferrer AB (2006) Decline in rice grain yields with temperature: models and correlations can give different estimates. Field Crop Res 98(2–3):151–156

Thompson LM (1986) Climatic-change; weather variability; and corn production. Agron J 78(4):649–653

Zhang T, Zhu J, Yang X, Zhang X (2008) Correlation changes between rice yields in North and Northwest China and ENSO from 1960 to 2004. Agric For Meteorol 148(6–7):1021–1033

Chapter 6
Crop Responses to Climate and Weather: Cross-Section and Panel Models

Wolfram Schlenker

Abstract Crop choices vary by climate, e.g., Florida specializes in citrus crops while Iowa specializes in corn and soybeans. The advantage of a cross-sectional analysis is that it incorporates how farmers adapt to existing difference in average climate conditions across space. A potential downfall is omitted variable bias. A panel analysis can overcome omitted variable bias by including fixed effects to capture all additive time-invariant influences, yet does not account for the same set of adaptation possibilities.

6.1 Introduction

Researchers might be interested in the relationship between temperature and yields for various reasons: (i) to forecast a yield at a given place in a given year under existing weather conditions; (ii) to simulate the effects of changes in average weather (i.e., climate) in the future. There is a clear distinction between the two. The former relies on the fact that farmers in a location have optimized their production process and adapted to the given climate. A historic time series at the specific location is sufficient to predict yields under various weather outcomes. Imagine a field in Iowa that has been in production for several years. If one were interested in predicting yields in that location, a good guess is to look at what happened to yields in previous years under various weather conditions and use that relationship in the forecast. This is an adequate procedure as farmers have to sow a crop before the weather is realized. For example, corn is usually sown in early spring in Iowa. There is no way to switch the crop in June if the weather turned warmer (or colder) than expected. The farmer is stuck with the crop that was initially chosen. While there are some possible adaptation measures even after the crop is planted (for example, increased use of irrigation or other inputs), the major decision has been made. Hence a farmer uses the existing distribution of possible weather outcomes when making the planting decision.

W. Schlenker
420 West 118th St, New York, NY 10027
email: wolfram.schlenker@columbia.edu

D. Lobell and M. Burke (eds.), *Climate Change and Food Security*,
Advances in Global Change Research 37, DOI 10.1007/978-90-481-2953-9_6,

The situation is quite different if the goal is to predict the impacts of changing climate conditions. If Iowa is to become permanently warmer, farmers might find it no longer optimal to grow the same corn variety but rather switch to a longer season variety instead. If it becomes significantly warmer, farmers might prefer to switch to an entirely new crop, e.g., cotton or citrus, two crops currently grown in warmer climates. Looking at past data at a given location does not incorporate switching to a different crop variety or entirely different crop species as the analysis keeps the crop variety and crop species fixed. Using historic yield data at the location of interest therefore might give an inaccurate prediction of what farmers would to do if the climate permanently changed.

6.2 Cross-Sectional Analysis

A cross-sectional analysis of a specific crop would incorporate how a farmer switches to other crop varieties of the same crop (e.g., corn varieties). The idea is to compare corn yields in Iowa with corn yields in warmer states like Arkansas. The problem is that there are other differences between Iowa and Arkansas besides differences in climate. For example, soil quality varies a great deal between states. A cross-sectional analysis would have to account for all covariates to correctly identify the effect of climate on corn yields.

If one is interested in how farmers switch crops with changing climates, a multinomial regression of crop choice on climatic variables, again accounting for all other confounding differences across climate zones explicitly, would identify such switching using cross-sectional data. In a multinomial regression, various crop choices are coded as separate categories. For example, outcome 1 could be growing maize, outcome 2 growing millet, and outcome 3 growing sorghum. A farmer will pick the most profitable crop for a given climate. Methodologically, each crop yield is modeled as a function of the climate variables as well as other controls and an error term. If the error terms follow an extreme value distribution, the probability for choosing each possible outcome has a closed form solution that is used in a multinomial logit regression. If the error terms are normal, the probability of choosing various alternatives has no closed form solution and can only be solved numerically (Maddala 1986). The multinomial logit technique has been applied to crop choices in South America by Seo and Mendelsohn (2008).

One can even go a step further and use farmland values as the dependent variable to implicitly incorporate crop switching without limiting the analysis to certain crop types. Farmland values reflect the value of land if it is put to its most profitable use, whatever that may be. An example might illustrate this point. New York City's Mayflower Hotel On the Park was a medium sized hotel on Central Park West at 61st Street. In the early 2000s it sold for an astonishing 400 million dollars. Why would anybody pay 400 million dollars for a medium-sized hotel? The first thing the new owner did was to knock down the old hotel and build a new luxury condominium (15 Central Park West) that reported apartment sales exceeding 1 billion dollars a few years later. Investors saw a higher value in demolishing the old hotel and putting the parcel to better use.

The idea behind farmland values is the same. If an investor were to look at a farm in Iowa and realize that growing corn is no longer optimal, but growing cotton is, she would be calculating the discounted amount of profit that could be made by farming cotton and should be willing to pay up to that price for the parcel. In equilibrium, the price of farmland, like the price of any type of land, should reflect the best use to which it can be put. A hedonic or Ricardian model hence uses a multivariate regression that estimates a statistical relationship between farmland values and climatic variables, as well as other controls (Mendelsohn et al. 1994).

Other controls are necessary as other variables that impact farm profitability will also be reflected in farmland values. If a farm has more productive soil with higher yields, the additional yield boost will be priced into the farmland. Ask yourself: if you could buy an acre of land that gives you 120 bushels/acre versus the average 100 bushels/acre, you could sell an extra 20 bushels at 3 dollars per bushel for an additional 60 dollars/year (assuming the production cost are the same between the two plots). Discounting an annuity of 60 dollars using a 5% discount rate would give a net present value of $60/0.05 = 1{,}200$. Since one parcel of land offers an additional discounted value of 1,200 more than the comparison plot at comparable cost, a rational market participant would be willing to pay up to 1,200 dollars extra.

It should be clear that there are some crucial assumptions underlying a hedonic analysis. First, the hedonic analysis assumes that prices are fixed (it is a partial equilibrium analysis). The above example assumes a constant corn price of 3 dollars per bushel. If climate change alters the productivity of entire regions, the demand and supply of various goods will change and prices will adjust accordingly. Second, there might be significant transaction cost in buying/selling farmland (capital as well as labor have to move), which might not give an efficient market outcome. Third, all other confounding variables that are correlated with climate have to be correctly accounted for, e.g., differences in soils, labor costs, or access to markets. If a variable that is correlated with climate is omitted from the analysis, its effect will incorrectly be attributed to climate.

Omitting uncorrelated variables would not bias the coefficient. This is the important fact that one uses in randomized experiments like a medical trial where half the sample is given a drug whose effect the researcher is testing. The control group is usually given a sugar pill. A test of the difference in the outcome of treated patients with the outcome in the control group gives the average treatment effect. While other controls like age or sex might impact the effectiveness of the drug, they will not bias the estimate of the average treatment effect. There should be as many people of various ages in both the treatment and control group if the sample size is large enough and people were assigned randomly to the treatment and control group. In other words the treatment and control group are "balanced." The problem is that in most cases of a cross-sectional analysis the treatment and control group are not balanced. Soil quality, access to markets, and agricultural institutions (extension service) all vary greatly among countries or even on a sub-country level.

Unfortunately, there is no direct test for omitted variables. One possible sensitivity check is to systematically include and exclude various control variables and see whether the coefficient of interest changes. Robust coefficients are reassuring in the following way: omitted variables would have to be correlated with the variable of interest but not the variables that are included and excluded from the analysis. This follows from the fact that the results did not change when various other controls were included and excluded. If the omitted variable were correlated with some of these other controls, the coefficient should have changed. For example, Schlenker et al. (2006) estimate a hedonic regression of farmland values in the Eastern United States on climatic variables as well as other controls (soil measures as well as socio-economic measures like per-capita income). When these other controls are included/excluded, the coefficients on the climatic variables remain robust. This is at least partially reassuring as any omitted variable that biases the coefficient would have to be correlated with climate, but be uncorrelated with various soil measures and socio-economic variables. While such variables might still exist, the set of possible candidates seems to be at least smaller.

The hedonic analysis has been extensively applied in World Bank studies using both farmland values as well as net revenue (a profit measure) as the dependent variable. Since farmland values are the discounted sum of all future net benefits that can be obtained from a piece of land if it is put to the best use, the two are closely related. It should, however, be noted that the cross-sectional analysis is linking farmland values or *average* profit measures to average weather variables. It is questionable to link profit *in one particular year* to average weather variables, as random weather outcomes would induce considerable noise and could severely bias the analysis. Citrus trees in California and Florida are usually highly profitable. Yet, in a year when there is a late freeze that kills the harvest and results in very low profits, linking profits from that particularly year to average weather outcomes where freezes only occur infrequently could be very misleading. The next section about panel models discusses whether annual profit measures can be linked to annual weather outcomes.

Since most countries have a narrower climate range than the United States, a cross-sectional analysis is impossible to estimate. Consider a country as small as Lichtenstein with a uniform climate that makes it impossible for a researcher to compare farms in warmer climates to farms in colder climates. A crucial requirement for any cross-sectional study is climate variation across space. One potential solution is to pool data from various countries (Seo and Mendelsohn 2007). On the other hand, the potential downside is that it becomes more difficult to account for all other confounding differences. Not only soils differ between countries but also institutional variables such as political stability and access to credit. Recall that omitting variables that impact farmers that are correlated with climate will bias the coefficients of the climatic variables. For example, if hotter countries were less politically stable and accordingly exhibited lower investments in agriculture, a cross-sectional analysis would wrongfully attribute these politically economy outcome to temperature differences if they were not accurately modeled. More recent studies have therefore sometimes broken down the analysis by agro-ecological zones with comparable farmland and included region or country fixed effects (Seo et al. 2008), which are further discussed in the next section.

6.3 Panel Analysis

A panel analysis recognizes that there are fundamental differences between spatial units of the analysis (e.g., countries) and that it is a difficult task to account for all these differences explicitly in a model. If these influences impact yields in an additive fashion and if they are time invariant, one can use fixed effects to capture them. In defense of the first assumption, the large majority of regression models use a linear specification and hence all factors are assumed to have an additive influence. Only a nonlinear model in parameters would capture non-additive factors. As long as a cross-sectional analysis uses a linear specification in the parameters, the assumption underlying a fixed effect model is no stronger than modeling each effect directly. A linear model in the parameters includes the case where variables are logged, as the logged variable still interacts linearly. The second assumption that all country-specific variables are time-invariant needs some pondering. Countries can change over time, get new governments, or even split in two. In such a case it is advisable to use fixed effects for each distinct temporal subset.

A fixed effect is a dummy or indicator variable that is set to one if observations from a group (country) are included and is set to zero otherwise. A panel requires at least two observations by country as otherwise the indicator variable would absorb all variation in that country. One cannot include a time-invariant variable in a panel model that uses fixed effects as this variable would be collinear with the fixed effects. For this reason one can for example not estimate a panel model of farmland values that uses average weather as explanatory variable combined with fixed effects. Average weather by definition is constant within a group and hence a linear multiple of the indicator variable.

It can be shown that a fixed effect model is equivalent to a joint group-specific demeaning of the dependent as well as all independent variables (Wooldridge 2001). If we subtract group specific averages from both the dependent and all independent variables and run a linear regression, the coefficients will be identical but the standard errors need to be adjusted for the difference in degrees of freedom. For example, a panel model that regresses country-level yields on average temperature during each growing season can be estimated in two ways. First, one can include a dummy for each country. If one also includes a constant, the dummy for one country has to be dropped to avoid perfect mulicolinearity. Second, one can subtract the average yield in each country from the yearly observations of yields and subtract the average climate in a country from each weather outcome in the country. If one were to then run a linear regression of the demeaned yields on the demeaned average weather without any country-specific dummy variables one would obtain identical regression coefficients as in the first specification. While this is a noteworthy statistical artifact, it also has an important interpretation. The regression uses deviations from country-specific averages to identify the parameter of interest. As discussed in further detail below, this is equivalent to fitting a regression line through each country where the slope is forced to be the same for each country but the intercept is allowed to vary by country. All countries are forced to exhibit the same sensitivity to weather fluctuations. For a model that allows for a distribution of weather sensitivities (i.e., distribution of regression slopes) the interested reader is referred to random coefficient models.

If we return to the initial discussion of this chapter, a panel uses variation in weather (i.e., year-to-year fluctuations in weather) as a source of identification and not differences in average weather (climate). Such a model will not incorporate any adaptation to systematic shifts in average weather. Some researchers therefore favor a random effects model which takes a weighted average of the within-group variation (fluctuations in weather) as well as the between-group variation (differences in average weather or climate). The interested reader is referred to any intermediate econometrics or statistics textbook. Intuitively, a random effects model does not include a separate dummy variable for each group but rather assumes that there might be a group-specific additive error term. This group-specific error term will capture time-invariant additive constants. Since omitted variables are included as a special expression in the error term, the estimated coefficients will suffer from an omitted variable bias similar to a cross-sectional analysis.

Panel data analyses have been used with profits (Deschênes and Greenstone 2007) and yields (Schlenker and Roberts 2009) as the dependent variable. The advantage of profits is that all crops can be aggregated into one single measure instead of modeling each crop separately. At the same time, there is a potential downfall with a profit or net revenue measure: most studies simply take the difference between total agricultural sales and production expenditures in a given year. Such an analysis neglects that the amount sold is different from the amount produced as most commodities are storable. In high-productivity years when yields are above normal, prices are low and farmers have an incentive to put part of the harvest into storage, i.e., the quantity sold is less than the quantity produced. On the flip side, when yields are below normal and prices are high, farmers have an incentive to sell part of the inventory that was harvested in previous years and hence the quantity sold is higher than the quantity produced. As a result, storage smoothes reported sales and makes them smaller than the full amount produced in good years and larger than the full amount in bad years. This will bias the estimated weather coefficients towards zero (Fisher et al. 2009). What one would need is an economic profit measure (value of production minus production cost) instead of the accounting measure (value of sales minus production cost).

In summary, the advantage of a panel is that one does not have to worry as much about omitted variable bias as the fixed effects capture all time-invariant variables. The downside is that a panel might measure something very different from a cross-sectional analysis, e.g., might capture various sets of adaptation possibilities. Some authors argue that the adaptation possibility is always greater in the cross-section, which would be in line with the Le Chatelier's principle that costs in a constrained system are higher than when constraints are relaxed in the long-run (e.g., fixed capital becomes obsolete). This is, however, not necessarily true in agriculture. Sometimes, adaptation possibilities are available in the short-run that could not be sustained forever. For example, a one-time drought might be mitigated by pumping groundwater. The aquifer recharge might be small enough that such groundwater pumping could not be sustained forever if droughts were to become more frequent. Finally, storage might bias a panel of agricultural profits as it smoothes sales between periods. It is an important omitted variable in a panel analysis that uses agricultural sales in a given year (quantity sold multiplied by price) instead of the value of all goods produced (quantity produced multiplied by price).

6.4 An Illustrative Example

Graph A of Fig. 6.1 displays crop yields in Lesotho in light grey squares and South Africa in black triangles for the years 1961–2000. The x-axis displays average temperature during the growing season, while the y-axis displays log yields (note that this variable can be negative as the log of 1 is 0, so any yield less than 1 ton/ha is a negative number). Lesotho is colder than South Africa as all grey squares lie to the left of the black triangles. Average yields are also lower is Lesotho than in South Africa as the average y-value of the grey squares is lower than the average height of the triangles.

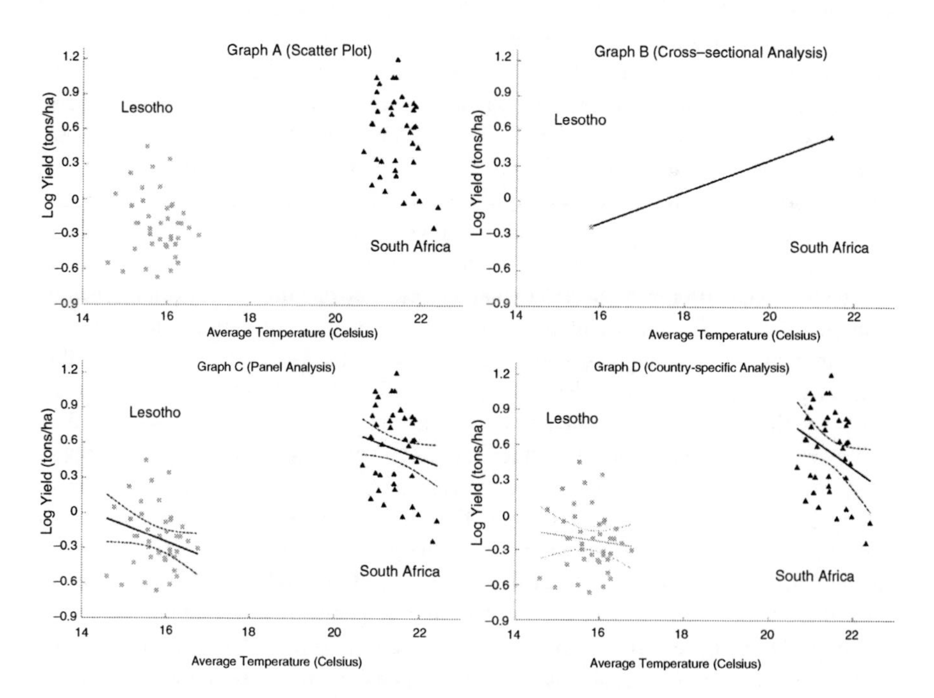

Fig. 6.1 Maize yields and average temperatures in Lesotho and South Africa (1961–2000)

Table 6.1 Summary statistics for maize in Lesotho and South Africa (1961–2000)

	Lesotho	South Africa
Average log yield (ton/ha)	–0.218	0.547
Average temperature (Celsius)	15.8	21.4
Fertilizer (kg/ha)	~1.8	21.7

A cross-sectional analysis would link average yields in Lesotho and South Africa to the respective average temperatures. This is done in graph B, where the 40 observations per country are reduced to 1 showing the average yield and average climate in a given country by averaging the 40 yearly observations. The solid line in panel B links the average outcomes in the two countries, which implies that increasing temperatures are beneficial as yields are increasing in average temperature.

One possible concern is that there are many differences between Lesotho and South Africa. Table 6.1 shows that South Africa uses much more fertilizer: 21.7 kg/ha compared to the 1.8 kg/ha in Lesotho. If each 1 kg/ha were to increase log yields by 0.0038 (i.e., increases yields by roughly 3.8%) than the entire difference in log yields ($0.765 = 0.547 + 0.218$) would be explained by the difference in fertilizer use ($19.8 = 21.7 - 1.8$) as $19.8 * 0.0038 \approx 0.765$ and the difference in temperatures would have no explanatory effect on the difference in yields. If the effect of fertilizer on yield is greater than 3.8% per kg, than the difference between South Africa and Lesotho *after adjusting for fertilizer use* implies that average yields are higher in Lesotho. If we link the yield net of fertilizer use again to temperatures, the estimated relationship would be negative as the country with higher average temperatures has lower average yields. Fertilizer is not the only difference between the two countries, and it should be immediately clear that it is empirically very challenging to account for all differences that might be correlated with differences in climate.

The intuition behind a panel analysis is shown in graph C of Fig. 6.1. It captures all time-invariant effects by an additive constant, i.e., a fixed effect. Differences in fertilizer use, institutional differences, access to markets, etc., are captured by a country-specific constant. This implies that each country can have a unique intercept of the regression line, but the slope of all regression lines is forced to be the same. Only variation within a country is used to identify the regression coefficient of interest (the slope of the regression line with respect to temperature): are yields in Lesotho and South Africa higher or lower in years that are warmer than usual? Note that the answer is the opposite of what we obtained in the cross-section (graph B): higher temperatures are worse, not better. However, analogous to the pure time series, it does not capture the full set of adaptation possibilities as we are identifying the parameter by looking at year-to-year fluctuations within a location. The difference between this and a time series is that we force these within country deviations to have the same effect among all countries.

Graph D of Fig. 6.1 shows the case of a pure time series which would estimate a separate regression for each country. The slope of the regression line is no longer the same but differs by country. A panel using fixed effects is closer to a time series model than to a cross-section. An alternative that uses both the variation within countries as well as the variation between country means (a weighted average of the time series and cross-section) is a random effects model (Wooldridge 2001).

Finally, a panel using fixed effects assumes the same effect of temperatures on yields in all countries, but the functional form decides whether this constant effect is in absolute or relative terms. In a linear model, a 1°C increase in average temperature is assumed to have the same *absolute* impact on yields, e.g., a decrease of 3 bushels/acre. In a log-linear model a 1°C is assumed to have the same *relative* impact on yields, e.g., a 3% decrease in yields. In a double-log model where both the dependent and independent variables are specified in logs, a *relative* temperature deviation (e.g., 5% less than normal) is assumed to have the same *relative* impact on yields, e.g., a 3% decline in yields.

6.5 A Brief Summary of Examples for the United States

The preceding sections have discussed the advantages and disadvantages of cross-sectional as well as panel models. While a cross-sectional analysis of farmland values seems appealing as it can capture crop switching, it might also suffer from omitted variable bias, which is less of a concern in a panel data set. While some argue strongly for cross-sectional analysis to measure adaptation (Mendelsohn et al. 1994), others will argue strongly for a panel model (Deschênes and Greenstone 2007) to avoid omitted variable bias. Since each model has its unique advantage, which model should be preferable?

The most fruitful exercise is to estimate various models and examine whether they agree or disagree. For example, in the case of US agriculture, the two models do agree if they are correctly specified and incorporate important agronomic principles like degree days. Schlenker et al. (2005) argue that highly subsidized irrigation water, which is correlated with climate and capitalizes into farmland values, is an important omitted variable in a cross-sectional analysis of farmland values in the entire United States and biases the climate coefficients. Schlenker et al. (2006) therefore estimate a model for the eastern United States only. While they include highly irrigated areas, e.g., 79% of the corn area in Arkansas was irrigated in 2007, they exclude farms with access to highly subsidized irrigation water in the Western United States. The highly subsidized public works programs in the West should not be counted as societal benefits but rather as a transfer from taxpayers to farmers. Moreover, the later analysis uses degree days instead of average temperatures. The resulting hedonic regression is highly stable between various Census years and gives robust estimates that very warm temperatures are the key drivers of farmland values.

Finally, a panel of crop yields reveals that the sensitivity of corn, soybeans, and cotton is comparable to the results of the hedonic regression in the sense that is predominantly extreme temperatures that determine yield outcomes (Schlenker and Roberts 2009). Moreover, the cross-sectional estimates for each crop are identical to panel and time series estimates, suggesting that there is limited potential for adaptation to extreme temperatures. In this sense, both cross-sectional analysis and panel models seem to give conforming answers.

6.6 Summary

This section has given a brief introduction to both cross-sectional as well as panel models. The advantage of a pure cross-section is that it can capture how farmers adapt to changing climatic conditions, which is very different from year-to-year weather fluctuations. The potential pitfall is that it might suffer from omitted variable bias as the coefficients of interest wrongfully pick up the effect of other variables that were incorrectly excluded from the analysis. A panel analysis offers a solution to avoid additive time-invariant omitted variables. However, it comes at a price: similar to time series models discussed in the proceeding section, such an analysis uses year-to-year weather fluctuations to identify the parameters of interest and hence might measure a very different set of possible adaptation measures. The difference between panel and time series models is that a panel forces the slope of the regression line to be constant for all groups (countries).

In general, it might be worthwhile to conduct all analyses and examine whether they differ. If they do differ, further analysis is necessary to resolve these differences.

References

Deschênes O, Greenstone M (2007) The economic impacts of climate change: evidence from agricultural output and random fluctuations in weather. Am Econ Rev 97(1):354–385

Fisher AC, Hanemamn WM, Roberts MJ, Schlenker W (2009) Agriculture and Climate Change Revisited. CUDARE Working Paper 1080

Maddala GS (1986) Limited-dependent and qualitative variables in econometrics. Cambridge University Press, UK

Mendelsohn R, Nordhaus WD, Shaw D (1994) The impact of global warming on agriculture: a ricardian analysis. Am Econ Rev 84(4):743–771

Schlenker W, Roberts M (2008) Estimating the Impact of Climate Change on Crop Yields: The Importance of Non-linear Temperature Effects. NBER Working Paper 13799

Schlenker W, Hanemann WM, Fisher AC (2005) Will US agriculture really benefit from global warming? Accounting for irrigation in the Hedonic approach. Am Econ Rev 95(1):395–406

Schlenker W, Hanemann WM, Fisher AC (2006) The impact of global warming on US agriculture: an Econometric analysis of optimal growing conditions. Rev Econ Stat 88(1):113–125

Seo SN, Mendelsohn R (2007) A Ricardian analysis of the impact of climate change on Latin American farms. World Bank Policy Research Working Paper 4163

Seo SN, Mendelsohn R (2008) An analysis of crop choice: adapting to climate change in South American farms. Ecol Econ 67(1):109–116

Seo SN, Mendelsohn R, Dinar A, Hassan R, Kurukulasuriya P (2008) A Ricardian analysis of the distribution of climate change impacts on agriculture across agro-ecological zones in Africa. World Bank Policy Research Working Paper 4599

Wooldridge JM (2001) Econometric analysis of cross section and panel data. MIT Press Cambridge, Mass

Chapter 7
Direct Effects of Rising Atmospheric Carbon Dioxide and Ozone on Crop Yields

Elizabeth A. Ainsworth and Justin M. McGrath

Abstract Rising atmospheric carbon dioxide concentration ([CO_2]) in this century will alter crop yield quantity and quality. It is important to understand the magnitude of the expected changes and the mechanisms involved in crop responses to elevated [CO_2] in order to adapt our food systems to the committed change in atmospheric [CO_2] and to accurately model future food supply. Free-Air CO_2 Enrichment (FACE) allows for crops to be grown in their production environment, under fully open air conditions, at elevated [CO_2]. Current best estimates for the response of the staple crops wheat, soybean and rice from FACE experiments are that grain yield will increase by 13% at 550 ppm CO_2. For the C_4 species, sorghum and maize, grain yield is not expected to increase at elevated [CO_2] if water supply is adequate. Grain quality is adversely affected by elevated [CO_2]. On average, protein content decreases by 10–14% in non-leguminous grain crops and concentrations of minerals, such as iron and zinc decrease by 15–30%. While these represent our best estimate of changes in crop yield quantity and quality, most studies have been done in temperate regions, and do not account for possible interactions of rising [CO_2] with other aspects of climate change, including increased temperature, drought stress and tropospheric ozone concentration.

7.1 Introduction

Carbon dioxide emissions from fossil–fuel burning and industrial processes have accelerated on the global scale over the past two decades (Canadell et al. 2007; Raupach et al. 2007). The growth rate of global atmospheric CO_2 for 2000–2006 was 1.93 ppm per year, which is the highest rate since the beginning of continuous

E.A. Ainsworth (✉)
USDA ARS Photosynthesis Research Unit; Department of Plant Biology, University of Illinois, Urbana-Champaign, 147 Edward R. Madigan Laboratory, 1201 W. Gregory Drive, Urbana, IL 61801, USA
email: Lisa.Ainsworth@ars.usda.gov

J.M. McGrath
USDA ARS Photosynthesis Research Unit

D. Lobell and M. Burke (eds.), *Climate Change and Food Security*,
Advances in Global Change Research 37, DOI 10.1007/978-90-481-2953-9_7,

monitoring in 1959. The current atmospheric CO_2 concentration ([CO_2]) of 385 ppm in 2008 (http://www.esrl.noaa.gov/gmd/ccgg/trends/) is higher than it has been in the past 650,000 years (Siegenthaler et al. 2005), and the concentration will continue to rise in the coming century. Atmospheric [CO_2] will likely reach 550 ppm by 2050 and 730–1020 ppm by 2100 (Meehl et al. 2007). Crops are therefore currently exposed to a [CO_2] that has not been experienced since the early Miocene, and agriculture is facing a future of uncertain consequences of global climate change.

Elevated [CO_2] directly stimulates photosynthesis in C_3 crops (e.g., wheat, rice and soybean), leading to increases in crop growth and seed yield (Kimball et al. 2002; Long et al. 2004; Nowak et al. 2004; Ainsworth and Long 2005). Elevated [CO_2] also directly decreases the conductance of CO_2 and water vapor through stomata, the pores in the leaf epidermis, of both C_3 and C_4 crops (e.g., maize, sorghum and sugarcane), which can improve water-use efficiency and therefore benefit all crop production in times and places of drought (Kimball et al. 2002; Ottman et al. 2001; Leakey et al. 2004, 2006; Leakey 2009). While rising [CO_2] is just one factor of global climate change, it plays a direct role in the sustainability of the future world food supply and projections of people at risk of hunger (Parry et al. 2004). Furthermore, the changes in temperature, precipitation and tropospheric ozone concentration projected for this century are spatially and temporally variable, while the increase in [CO_2] is uniform, global and committed (Solomon et al. 2007). Therefore, understanding crop responses to [CO_2] is a critical first step in adapting agriculture to anticipated global change.

In this chapter, we review the experimental approaches that have been used to investigate crop responses to rising atmospheric [CO_2], summarize the current understanding of how rising atmospheric [CO_2] will alter crop physiology and yield, discuss how models extrapolate this information beyond experimental settings to make predictions of food production and security in the future, discuss potential effects of elevated ozone, and identify major knowledge gaps and challenges for future research.

7.2 Experimental Approaches for Investigating Crop Production in Elevated [CO_2]

Several technologies have been used to study the effects of elevated [CO_2] on crop productivity, including controlled environmental chambers, greenhouses, open-top chambers (OTC), and Free-Air CO_2 Enrichment (FACE; Long et al. 2004). In controlled environmental chambers and greenhouses, plants are typically grown in pots, with lighting, nutrients and water supplied by the researcher in specified amounts. There are practical advantages to using controlled environments, including precise control of precipitation, humidity and light, as well as ready availability of such facilities at academic and government research laboratories. Controlled environments have also been used to conduct dose response curves for crops grown at a range of elevated [CO_2] (e.g., Allen et al. 1987; Long et al. 2006). However, there are several drawbacks to use of these technologies in the context of investigating crop yield responses to climate change (Long et al. 2004). Pots limit root growth, which can negatively feed back on photosynthetic capacity, shoot growth and harvestable yield potential, and thus reduce the magnitude of CO_2 stimulation (Arp 1991). Growth in pots can also alter nutrient availability, thereby changing the CO_2 response (McConnaughay et al. 1993). The size, light levels and forced air circulation in controlled environments also alter plant growth, which compromises the ability to accurately measure crop yield responses to [CO_2] (McLeod and Long 1999; Long et al. 2004, 2006).

In the field, crops can be grown in OTCs, where plants are rooted in the ground and exposed to natural light and precipitation through the top of the chamber (Heagle et al. 1989; Leadley and Drake 1993; Whitehead et al. 1995). OTC walls are typically clear plastic, allowing light penetration, and air enriched with CO_2 is introduced to the chamber by a blower system. Although OTCs eliminate some of the problems associated with greenhouses and growth chambers, OTCs alter the environmental conditions, such that temperatures and relative humidity are higher, wind velocity and light intensity are lower, and light quality is changed (Leadley and Drake 1993). Another problem with OTCs is their small plot size. Typically, agronomic trials also use buffer rows, with a width approximately twice the height of the crop. However, with OTCs, most of the treated crop is within the buffer zone, which causes "edge effects" and could exaggerate the response to elevated [CO_2] (McLeod and Long 1999).

In response to the limitations of controlled environments and OTCs and as the need to test hypotheses under open-air field conditions arose, FACE technology was developed (Hendrey and Miglietta 2006; Fig. 7.1). FACE allows elevated [CO_2] to be maintained without significantly altering the micrometeorological conditions around a plot of vegetation (Hendrey et al. 1993). FACE plots encompass up to hundreds of square meters of vegetation, allowing for use of a buffer zone, which eliminates problems of edge effects experienced in chambers (Long et al. 2006). The size of FACE plots also enables investigation of plant responses to

Fig. 7.1 A FACE plot at the University of Illinois SoyFACE facility where soybean is exposed to elevated [CO_2] (550 ppm). CO_2 is released from small holes in the green pipe into the wind, on the upwind side of the plot. The release rate is determined by the wind speed and [CO_2], measured at the center of the ring (photo credit: Andrew D.B. Leakey)

elevated [CO_2] from the genomic to ecological scale (Leakey et al. 2009b). FACE systems release CO_2-enriched air through vertical vent pipes (e.g., Lewin et al. 1992) or pure CO_2 through horizontal pipes (Miglietta et al. 2001). The gas is released just above the canopy surface on the upwind side of the plot. Fast-feedback computer control adjusts the position and amount of CO_2 released at different points around the plot, based on measurements of wind speed, direction and [CO_2] in the center of the plot (Long et al. 2004). In FACE plots, the natural environment is essentially unperturbed, as there are no barriers to light, precipitation, wind or pests. A major limitation to widespread use of FACE experimentation is the financial investment in the infrastructure, land and personnel needed to successfully run the experiments (Ainsworth et al. 2008a). Therefore, far fewer FACE experiments have been conducted than controlled environment studies. Still, because plants are grown in soil without significant alteration of the microenvironment, FACE experiments likely offer the most realistic estimates of crop yield responses to elevated [CO_2] (Long et al. 2004, 2006; Ainsworth et al. 2008b).

7.3 Direct Effects of Elevated [CO_2] on Plant Physiology

There are two direct, instantaneous effects of elevated [CO_2] on C_3 plants: an increase in photosynthetic carbon gain and a decrease in stomatal conductance of CO_2 and water vapor. Any stimulation of crop yield by elevated [CO_2] is principally determined by those two fundamental responses (Farquhar et al. 1978; Drake et al. 1997; Long 1999; Long et al. 2004; Ainsworth and Rogers 2007). An immediate rise in [CO_2] increases the net photosynthetic carbon gain in C_3 plants because ribulose-1,5-bisphosphate carboxylase–oxygenase (Rubisco), the enzyme that initially fixes CO_2, is not saturated in today's atmosphere. Therefore, the velocity of Rubisco carboxylation reactions increases with rising [CO_2]. Rising [CO_2] also competitively inhibits the oxygenation reaction, which improves the efficiency of net carbon gain by decreasing photorespiratory CO_2 loss (Bowes 1991; Long et al. 2004).

The immediate gains in photosynthesis are not always maintained at the same magnitude when plants are grown at elevated [CO_2] for longer durations. Growth at elevated [CO_2] results in altered photosynthetic capacity in C_3 crops, namely decreased maximum Rubisco activity (Drake et al. 1997; Long et al. 2004; Ainsworth and Long 2005). This mechanism is thought to operate to optimize utilization of nitrogen, and on average, C_3 crops show a 17% decrease in maximum Rubisco activity when grown at 567 ppm, based on the results of recent FACE experiments (Ainsworth and Rogers 2007). Environmental and genetic factors that limit sink strength (e.g., grain number) and lead to accumulation of carbohydrate content in leaves are associated with down-regulation of photosynthetic capacity (Long et al. 2004; Ainsworth and Rogers 2007; Leakey et al. 2009a). For example, when isogenic lines of soybean (*Glycine max*) were exposed to elevated [CO_2], only the non-nodulating line showed decreased photosynthetic capacity (Ainsworth et al. 2004). Low nitrogen fertilization exacerbates any shortage of nitrogen relative to carbon and results in significant and pronounced decreases in photosynthetic capacity of C_3 crops (Ainsworth and Long 2005).

However, despite the changes in photosynthetic capacity, carbon gain is significantly greater in C_3 plants grown at elevated [CO_2] anticipated for the middle to end of this century. On average, daily photosynthetic carbon gain increased by 9% for rice (*Oryza sativa*), 13% for wheat (*Triticum aestivum*) and 19% for soybean grown at elevated [CO_2] in FACE experiments (Long et al. 2006). The increase in carbon gain in C_3 crops feeds forward to increased vegetative and reproductive growth, and harvestable yield (Ainsworth and Long 2005; Long et al. 2006).

The second direct effect of elevated [CO_2] on plants is decreased stomatal conductance of CO_2 and water vapor (Long et al. 2004; Ainsworth and Rogers 2007). Decreased stomatal conductance is common to both C_3 and C_4 species, unlike the direct stimulation of photosynthesis, which is only observed in C_3 species. C_4 species concentrate CO_2 in bundle sheath cells where Rubisco is located, which essentially saturates the carboxylation reaction and eliminates photorespiration in C_4 species (von Caemmerer and Furbank 2003). However, both C_3 and C_4 species show decreased stomatal conductance at elevated [CO_2]. While a change in stomatal conductance does not always translate into an equivalent change in canopy water use, recent FACE experiments with both C_3 and C_4 crops reported 5–20% reductions in canopy transpiration (reviewed in Leakey et al. 2009a). Changes in canopy transpiration at elevated [CO_2] were also associated with improvements in soil moisture content (Conley et al. 2001; Hunsaker et al. 2000; Leakey et al. 2006), and maintenance of canopy carbon gain during dry periods (Leakey et al. 2004; Bernacchi et al. 2007). The direct effect of elevated [CO_2] on stomatal conductance provides a second means for improvement of both C_3 and C_4 crop yield at elevated [CO_2] in times and places of drought (Leakey 2009).

7.4 Crop Yield Responses to Elevated [CO_2]

The direct effects of elevated [CO_2] on photosynthesis and stomatal conductance lead to changes in crop growth, carbon allocation, biomass accumulation and ultimately seed yield. It is well established that stimulation of seed yield by elevated [CO_2] is lower in magnitude than stimulation of photosynthesis and above-ground biomass, suggesting that feedbacks constrain the potential benefits of elevated [CO_2] (Long et al. 2004). For example, in soybean, night-time foliar respiration is stimulated by elevated [CO_2] (Leakey et al. 2009c), which reduces plant carbon balance, but may be necessary to produce energy for export of additional carbohydrate from leaves to reproductive sinks. In addition, there may be bottlenecks that limit transport of fixed carbon into economic yield that should be targets for further study (Ainsworth et al. 2008c). The following sections describe what we know about the magnitude of crop seed yield responses to elevated [CO_2], the mechanisms for those changes, and how yield quality is altered by elevated [CO_2].

7.4.1 Changes in Yield Quantity at Elevated [CO_2]

The large number of experiments in controlled environments has allowed yield dose response curves to be calculated for the major C_3 food crops, soybean, wheat and rice. In Fig. 7.2, the ratio of yield at elevated [CO_2] relative to ambient [CO_2] was calculated from all available studies of soybean, wheat and rice grown to maturity at elevated [CO_2] in controlled environments and open-top chambers (for original references, see Ainsworth et al. 2002 for soybean, Amthor 2001 for wheat, Ainsworth 2008 for rice). After averaging all studies within 100 ppm intervals, a non-rectangular hyperbola was fit to the data independently for each crop (as in Long et al. 2006; Fig. 7.2). The response of C_3 crop yield to [CO_2] is approximately hyperbolic, increasing linearly at sub-ambient, and saturating at approximately 1,000 ppm (Fig. 7.2). This theoretical response of crop yield to [CO_2] is expected based on the response of photosynthetic carbon gain to elevated [CO_2] (Allen et al. 1987).

Although an increase in C_3 crop yield to elevated [CO_2] is supported by controlled environment, OTC and FACE studies, the magnitude of the change in yield has been the subject of ongoing debate in the literature (Long et al. 2006; Tubiello et al. 2007a, b; Ainsworth et al. 2008b). Superimposed upon each yield response curve in Fig. 7.2 is the average yield response to elevated [CO_2] from the FACE experiments (open symbols). The stimulation of yield at elevated [CO_2] (~550 ppm) observed in FACE experiments is approximately half the stimulation predicted from the controlled environment studies (Fig. 7.2). Rice, wheat and soybean yields were increased by 12, 13 and 14% respectively by growth at elevated [CO_2] in FACE (Long et al. 2006), compared to an approximate 30% increase for those crops at 550 ppm predicted by the hyperbolic yield response curves.

A limitation in this comparison is that FACE experiments have primarily used a single elevated [CO_2], close to 550 ppm, the concentration anticipated for the middle of this century. FACE experiments have also been conducted over a narrower range of ambient [CO_2] compared to controlled environment studies. Therefore, a more direct comparison of yield results from controlled environments and FACE was taken by limiting the comparison of FACE experiments and chamber studies to those with similar ambient [CO_2] and similar elevated [CO_2] (Ainsworth 2008; Ainsworth et al. 2008b). This more direct comparison showed a wider range of responses of C_3 crops to elevated [CO_2] in controlled environment studies, but confirmed the result that the stimulation in harvestable yield at elevated [CO_2] in FACE experiments is approximately half of the stimulation in controlled environments (Ainsworth et al. 2008b).

A number of C_3 crops other than staple cereals have been grown at elevated [CO_2] in FACE experiments, including potato (*Solanum tuberosum*), barley (*Hordeum vulgare*), sugar beet (*Beta vulgaris*) and oilseed rape (*Brassica napus*).

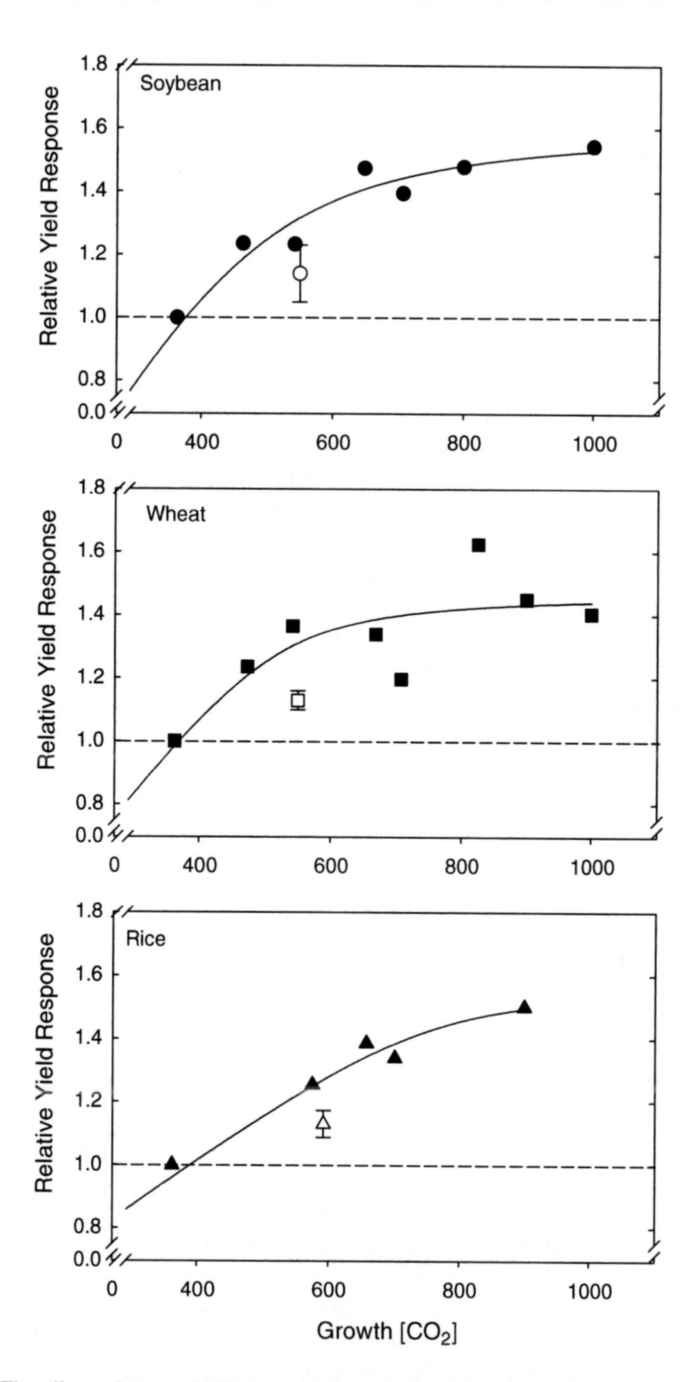

Fig. 7.2 The effects of elevated [CO_2] on soybean, wheat and rice yield (adapted from Long et al. 2006; Ainsworth 2008). Data are yields at elevated [CO_2] relative to yield at ambient [CO_2] for crops grown in enclosures (*solid symbols*) and FACE (*open symbols*). *Error bars* represent 90% confidence intervals around the means for the FACE studies. The *solid lines* are the least squares fit for the nonrectangular hyperbolic response of yield to growth [CO_2] from the enclosure studies of soybean ($r^2=0.98$), wheat ($r^2=0.88$) and rice ($r^2=0.96$)

Two cultivars of potato were grown at elevated [CO_2] in Central Italy (Miglietta et al. 1998; Bindi et al. 1999). In the first experiment, the Primura cultivar was exposed to three elevated [CO_2]: 460, 560 and 660 ppm. Tuber dry mass was stimulated by 13.8, 27.7 and 41.5% at each respective [CO_2] (Bindi et al. 2006). In the second experiment with potato, the Bintje cultivar was grown at 550 ppm for two growing seasons and showed a 36–50% increase in tuber dry mass (Bindi et al. 2006). These results suggest that other tuber crops, such as cassava and sweet potato, also have the potential to benefit from elevated [CO_2]; however, when these crops are grown with limited fertilization and under water stress, gains may not be realized.

Sugar beet and barley were grown at elevated [CO_2] (550 ppm) in a FACE experiment in Braunschweig, Germany (Weigel et al. 2006). Sugar beet showed a 6–8% stimulation in tuber production at elevated [CO_2], while barley showed an 8–14% stimulation in grain production (Weigel et al. 2006). Oilseed rape yield was ~18% higher when grown at elevated [CO_2] of 500 ppm (Franzaring et al. 2008).

The response of two C_4 crops to elevated [CO_2] has been studied in FACE experiments (Ottman et al. 2001; Leakey et al. 2004, 2006). Sorghum (*Sorgum bicolor*) was grown in Maricopa, Arizona with and without ample water supply, and maize (*Zea mays*) was grown in Champaign, Illinois with ambient precipitation. Elevated [CO_2] had no effect on seed yield when averaged across growth conditions and two growing seasons for each crop (Ottman et al. 2001; Leakey et al. 2004, 2006). There was a trend towards an increase in sorghum yield when the crop was grown without ample water supply (Ottman et al. 2001). While millets and sugarcane (*Saccharum* sp.) have not been grown at elevated [CO_2] in FACE experiments, a recent OTC study of sugarcane revealed that elevated [CO_2] (720 ppm) increased photosynthesis by 30%, height by 17%, biomass by 40% and sucrose content by 29% (De Souza et al. 2008). These data suggest that sugarcane productivity might increase in the future; however, the OTCs also may have overestimated the effects of elevated [CO_2] and caused transient water stress (De Souza et al. 2008).

7.4.2 Change in Components of Yield at Elevated [CO_2]

The consistent stimulation of economic yield at elevated [CO_2] can involve larger seed or grain size, more seeds per pod, ear or panicle, and/or more reproductive structures per plant. The yield benefit for most C_3 crops resulted from increased above-ground dry matter production supporting more reproductive structures (Table 7.1). In the FACE experiments with rice, tiller, panicle and spikelet numbers per area increased significantly (Kim et al. 2003; Yang et al. 2006). Those increases in dry matter production were large enough to outweigh negative effects of elevated [CO_2] on productive tiller ratio and degenerated spikelets (Kim et al. 2003; Yang et al. 2006).

In C_3 oilseed and grain crops, elevated [CO_2] also had little effect on individual grain or seed mass (Table 7.1). This may not be unexpected since individual grain weight has not changed with genetic improvement in wheat, rice, or soybean over much of the last century (Morrison et al. 2000; Fischer 2007). All C_3 oilseed and

Table 7.1 Average percent change in economic yield, final above-ground biomass, individual seed or grain weight, and harvest index of crops grown at elevated [CO_2] (~550 ppm) in FACE experiments. Bold numbers represent statistically significant changes ($p<0.10$) reported in primary literature sources

Crop	Economic yield (%)	Above-ground biomass (%)	Individual seed or grain weight (%)	Harvest index (%)
Soybean[a]	**+14**	**+16**	0	**−2**
Wheat[b]	**+13**	**+10**	–	–
Rice[c]	**+13**	**+27**	+1	−2
Potato[d]	**+34**	−5	–	–
Oilseed rape[e]	+18	+17	+18	−3
Maize[f]	0	−2	−1	−2
Sorghum[g]	+4	**+9**	−1	−2

[a]Morgan et al. 2005; [b]Pinter et al. 1996, Kimball 2006; [c]Data are averaged from the following studies: Kim et al. 2001, 2003; Pang et al. 2006; Sasaki et al. 2005; Seneweera et al. 2002; Shimono et al. 2007; Yang et al. 2006; [d]Bindi et al. 2006; [e]Franzaring et al. 2008; [f]Leakey et al. 2006; [g]Ottman et al. 2001.

grain crops had a lower harvest index at elevated [CO_2]; however, the magnitude of the change was only significant for soybean and the change in all crops was small (Table 7.1). Still, maintaining current levels of harvest index represents one potential area for improving crop responses to elevated [CO_2] (Ainsworth et al. 2008c).

Potato differed from the other C_3 crops (Table 7.1). Tuber production was significantly increased by elevated [CO_2] while aboveground dry matter production was not affected (Bindi et al. 2006). The number of tubers, rather than the size of the tubers, caused the enhancement in yield (Miglietta et al. 1998). Also, the fraction of malformed tubers was not affected by elevated [CO_2] (Bindi et al. 2006).

When averaged across all experiments, final yield, grain weight and harvest index of C_4 crops was not affected by growth at elevated [CO_2] (Table 7.1). When sorghum was grown at elevated [CO_2] under conditions with water stress, there was a tendency towards higher yields and greater aboveground biomass (Ottman et al. 2001). This supports the notion that C_4 plants will benefit from elevated [CO_2] in times and places with drought, but more studies are needed to reduce uncertainty in this prediction (Leakey 2009).

7.4.3 *Changes in Yield Quality at Elevated [CO_2]*

Much of the focus on the effects of elevated [CO_2] on crops has been on harvestable yield quantity. However, yield quality is an important issue as well. The two most studied aspects of quality are protein and nitrogen concentration. A meta-analysis of crops grown in elevated [CO_2] found that protein content was reduced in grain (Taub et al. 2008). Barley, wheat, rice, potato and soybean all showed significant decreases; for the non-legumes, the decrease was between 10% and 14%, whereas for soybean, a legume, the decrease was a much smaller 1.5% (Taub et al. 2008). This is likely because legumes are able to fix nitrogen, which would prevent nitrogen dilution. Nitrogen concentration was also decreased in the grains of wheat (Kimball 2006, Manderscheid et al. 1995), barley (Manderscheid et al. 1995) and rice (Kobayashi et al. 2006).

Change in mineral quality of the harvestable portion of crops has been less extensively studied although the data suggest that mineral content is generally reduced by growth in elevated [CO_2] (Loladze 2002). In two wheat and two barley cultivars grown at 718 compared to 384 ppm [CO_2], several minerals including Ca, Mg, and S were reduced between 2% and 20%, and Fe and Zn were decreased by 15–20%, with larger decreases in wheat than barley (Manderschied et al. 1995). In one wheat cultivar grown at 550 compared to 380 ppm [CO_2], Ca, S and Fe decreased by 10%, but P, K, Zn and Mn were unchanged (Fangmeier et al. 1999). In rice grown at 700 compared to 370 ppm [CO_2], across a range of P concentrations, Fe was reduced by up to 30% and Zn by up to 14.5%, while Ca was increased by 5% (Seneweera and Conroy 1997).

Changes in protein and mineral content in grains have significant consequences for animal and human health. Livestock that are deficient in certain minerals have decreased fertility and productivity even if the deficiency is not great enough for the animal to present clinical symptoms (Fisher 2008). Therefore, in future atmospheric conditions, if animal feed is not supplemented with minerals (an option not available to all farmers) then production of animal-based food and products (e.g., wool) might be reduced compared to production in current [CO_2]. Reductions in Fe and Zn content would also have important direct consequences for humans, considering the large numbers of people that currently suffer from micronutrient deficiency (see Chapter 2). Although the risk of Fe deficiency is greatest in developing counties (56% of pregnant women and 54% of school-age children), it is also a large problem in developed countries (18% of pregnant women and 17% of school-age children; UN ACC/SCN 2000).

Although the decrease in protein could be reduced by addition of nitrogen fertilizer, the cost and availability of fertilizer prohibits widespread adoption by all farmers, particularly in developing countries. Furthermore, additional fertilizer use has a significant environmental cost. Therefore, the decrease in protein content in grains will likely be a problem in the future (Taub et al. 2008). Similarly, while nutrient deficiency can be avoided by eating a more varied diet, taking dietary supplements, augmenting commercial food products with nutrients (a process termed fortification), or breeding crops with increased nutrient content, not all people have access to more varied foods or nutrient supplements. There has been some success in genetically improving nutrient content in crops. For example, the maize line, Opaque 2, has been bred to produce 32% more lysine, an essential amino acid that is typically found in very low concentrations in maize (Higgins and Chrispeels 2003). However, access to enhanced germplasm is not widely available to farmers in developing countries where these deficiencies pose the greatest risk to the population.

7.5 Modeling the Impact of [CO_2] on Crop Production and Global Food Supply

The response of crop yields to elevated [CO_2] is a key parameter in projections of the effects of climate change on global crop yields, world food supply and risk of hunger in the future (e.g., Parry et al. 1999, 2004). Inclusion of the direct effects of elevated [CO_2] in a recent assessment substantially improved estimated world cereal prices and reduced the risk of hunger for 500 million people by 2080 (Parry et al. 2005). Process-based crop models with deterministic equations of underlying physiological processes compute crop growth and development, biomass partitioning and economic yield in response to environmental inputs (see Chapter 4). A CO_2-response factor can be applied to different physiological processes in the models in order to reflect the direct or indirect effects of elevated [CO_2] (reviewed by Tubiello and Ewert 2002). While the direct, instantaneous effects of elevated [CO_2] on photosynthesis and stomatal conductance can be accurately modeled (Farquhar et al. 1980; Ball et al. 1987), scaling these direct effects into long-term crop growth and ultimately seed yield is much more challenging.

Furthermore, mechanistic equations such as the Farquhar et al. (1980) model of C_3 photosynthesis are only occasionally used to calculate a CO_2 response (e.g., in the DEMETER model, Kartschall et al. 1995); more often, simple linear or curvi-linear multipliers are used to model the effects of [CO_2] on photosynthesis, stomatal conductance, carbon partitioning, plant water relations and/or yield (Tubiello and Ewert 2002; Parry et al. 2004). Literature reviews from the 1980s are reportedly used as the basis for these linear or curvi-linear multipliers (e.g., Kimball 1983; Rogers et al. 1983; Cure and Acock 1986; Allen et al. 1987; Peart et al. 1989), which has raised the concern that estimates of future food supply may be overly optimistic (Long et al. 2005, 2006; Ainsworth et al. 2008b). While this issue has been the subject of debate, it is clear that before incorporating any crop model into an assessment of climate change impacts on global crop production and food supply, it is critical to evaluate the crop model's performance against field data (Tubiello and Ewert 2002).

A number of process-based crop models have been evaluated against data from FACE experiments (Tubiello et al. 1999; Ewert et al. 2002; Kartschall et al. 1995; Grossman-Clarke et al. 2001; Grant et al. 1999; Jamieson et al. 2000; Bannayan et al. 2005; Asseng et al. 2004). In addition to CO_2 treatments, these models have tested the interaction of CO_2 with drought stress and N supply. A comparison of effects of elevated [CO_2] on wheat and rice grain yield from two FACE experiments and five crop model simulations is shown in Fig. 7.3. LINTULCC2 and AFRCWHEAT2 were able to capture the stronger effect of elevated [CO_2] on wheat yields under water-stressed conditions compared to well-watered conditions; however, the magnitude of the stimulation in the model was greater than the stimulation in the field (Fig. 7.3a; Ewert et al. 2002). APSIM-N also captured a positive effect of elevated [CO_2] on wheat yield under high N fertilization, but again the magnitude of the modeled response was greater than the FACE result (Fig. 7.3b; Asseng et al. 2004). Finally, the Oryza2000 model matched very well with the FACE results,

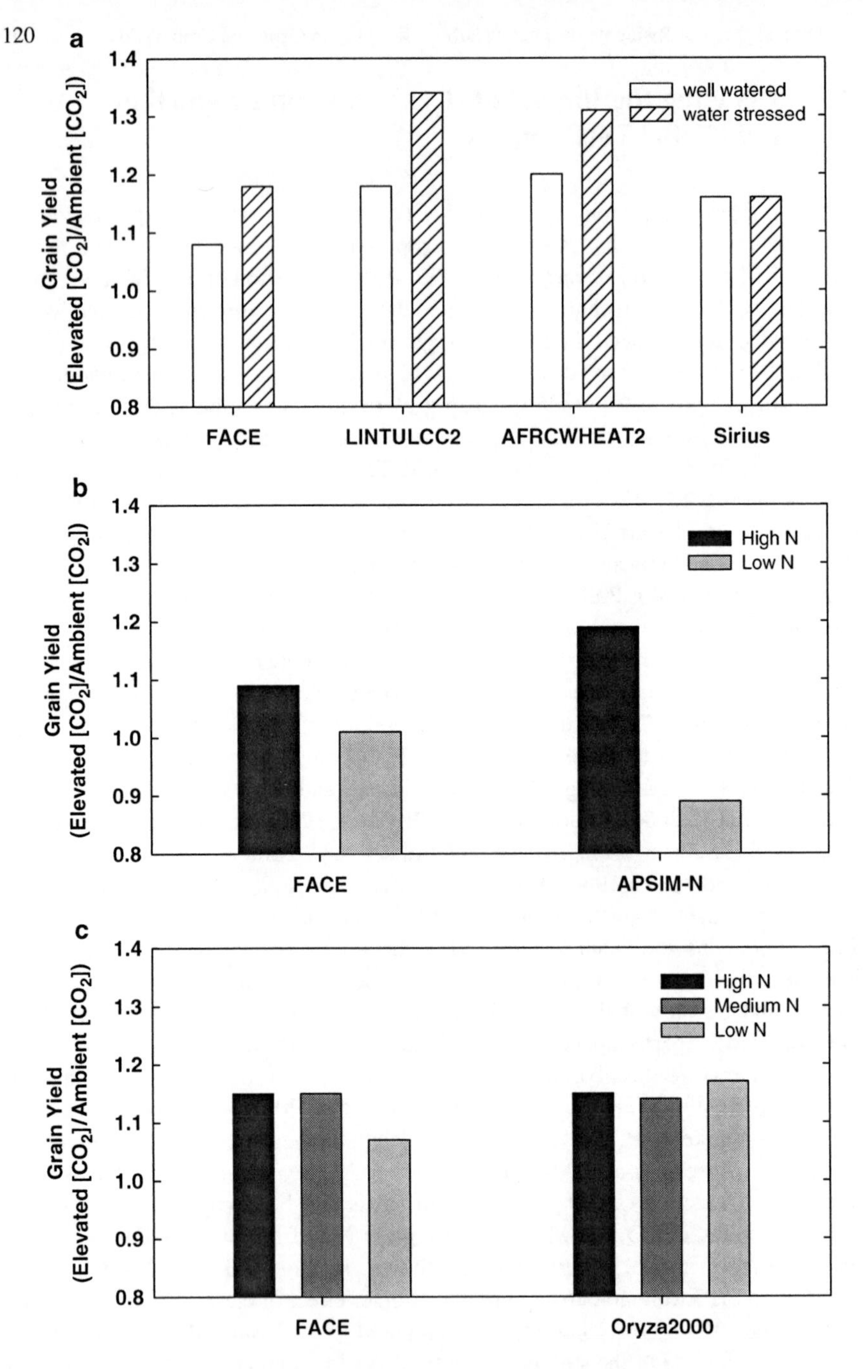

Fig. 7.3 A comparison of stimulation of crop yields from crop models and FACE experiments. (**a**) A comparison of wheat yields from three crop models (LINTULCC2, AFRCWHEAT2, Sirius) with the Maricopa Free-Air CO_2 Enrichment (FACE) experiment (Ewert et al. 2002). (**b**) A comparison of the APSIM-N crop model with wheat yield results from the Maricopa FACE experiment (Asseng et al. 2004). (**c**) A comparison of the Oryza2000 crop model with rice yield results from a FACE experiment in northern Japan (Bannayan et al. 2005)

except for under low N conditions (Fig. 7.3c; Bannayan et al. 2005). Overall, the models successfully captured the direction of the response of wheat and rice yields to elevated [CO_2], but the magnitude of the modeled output was often significantly different from the experimental results (Fig. 7.3). More often than not, the crop models overestimated the actual yield stimulation measured in the field.

7.6 Crop Responses to Elevated Ozone

Tropospheric ozone concentrations ([O_3]) have more than doubled over land in the Northern Hemisphere since pre-industrial times (Akimoto 2003; Vingarzan 2004). Ozone is a dynamic secondary pollutant formed from the photochemical oxidation of methane, carbon monoxide, and volatile organic compounds in the presence of nitrogen oxides. Hot, sunny weather favors formation of ozone in the troposphere, and high concentrations can occur across large areas, far from industrial sources (Ashmore 2005). Between 1876 and 1910, background O_3 concentrations were estimated to range from 5 to 16 ppb (Volz and Kley 1988), while modern day global annual mean O_3 concentrations range from approximately 28 ppb in South America to 45 ppb in Southern Asia (Dentener et al. 2006). Unlike CO_2, which is relatively well mixed in the atmosphere, there is significant variability in [O_3] depending on geographic location, elevation and the extent of anthropogenic sources (Vingarzan 2004; Ashmore 2005). In the major crop growing regions of the United States in 2005, the daytime surface O_3 concentrations during summer months ranged from 50 to 65 ppb (Tong et al. 2007). The future [O_3] will depend upon anthropogenic emissions, trends in temperature, humidity and solar radiation, and implementation of air quality legislation. Only with a global implementation of O_3 precursor control measures will background [O_3] decrease in the future. Without rapid and global implementation of legislation, by 2030 average [O_3] over the Northern Hemisphere could increase by 2 to 7 ppb, and by 2100, extreme emission scenarios project a baseline increase of more than 20 ppb (Prather et al. 2003).

Ozone enters plants through the stomata, where it reacts to form other reactive oxygen species, which in turn alter a number of physiological processes (Fiscus et al. 2005; Fuhrer 2009). Ozone decreases photosynthetic carbon gain by impairing Rubisco activity and reducing stomatal conductance (e.g., Morgan et al. 2004), inhibits reproduction by affecting pollen germination, fertilization and abortion of flowers (Black et al. 2000), impairs phloem loading and assimilate partitioning to roots and grains (Fuhrer and Booker, 2003), and decreases aboveground biomass, individual grain number and mass, and final harvestable yield (Morgan et al. 2003; Ainsworth 2008; Feng et al. 2008).

A number of different exposure indicators are used to calculate dose response functions, including seasonal 7 and 12 h mean [O_3] during daylight, and seasonal cumulative exposure over a threshold of 40 ppb (AOT40) or 60 ppb (SUM06) (Mauzerall and Wang 2001). Crop-specific O_3-exposure functions, which relate a quantifiable O_3-exposure indicator to reductions in crop yield, have been developed from extensive OTC studies in the United States (National Crop Loss Assessment Network – NCLAN) and Europe (European Open Top Chamber Program – EOTCP)

(Heck et al. 1987; Fuhrer et al. 1997), and are used to assess both current and future levels of crop loss to O_3. Mills et al. (2007) synthesized linear AOT40-based response functions for different crops from over 700 studies, and found that there were three significantly different groups of responses (Fig. 7.4). Wheat, watermelon, pulses, cotton, turnip, tomato, onion, soybean and lettuce were O_3-sensitive; sugar beet, potato, oilseed rape, tobacco, rice, maize, grape and broccoli were moderately sensitive; and barley, plum and strawberry were O_3-resistant (Fig. 7.4).

Ozone-crop yield response functions can be used with different emissions scenarios and global chemistry transport models to estimate current and future relative yield losses to [O_3] (e.g., Wang and Mauzerall 2004; Tong et al. 2007; Van Dingenen et al. 2009). Using the IPCC B2 scenario of moderate population growth, intermediate levels of economic development and increased concern for environmental and social sustainability, Wang and Mauzerall (2004) projected that between 1990 and 2020, grain yield loss to [O_3] would increase by 35, 65 and 85% in Japan, Korea and China, respectively. In a global analysis with the optimistic scenario that all current emissions legislation will be fully implemented by 2030, Van Dingenen et al. (2009) project that the global relative yield losses to O_3 will increase by 4% for wheat, 0.5% for soybean, 0.2% for maize and 1.7% for rice by 2030. Clearly, estimates of the effects of O_3 on future crop production depend upon trends in

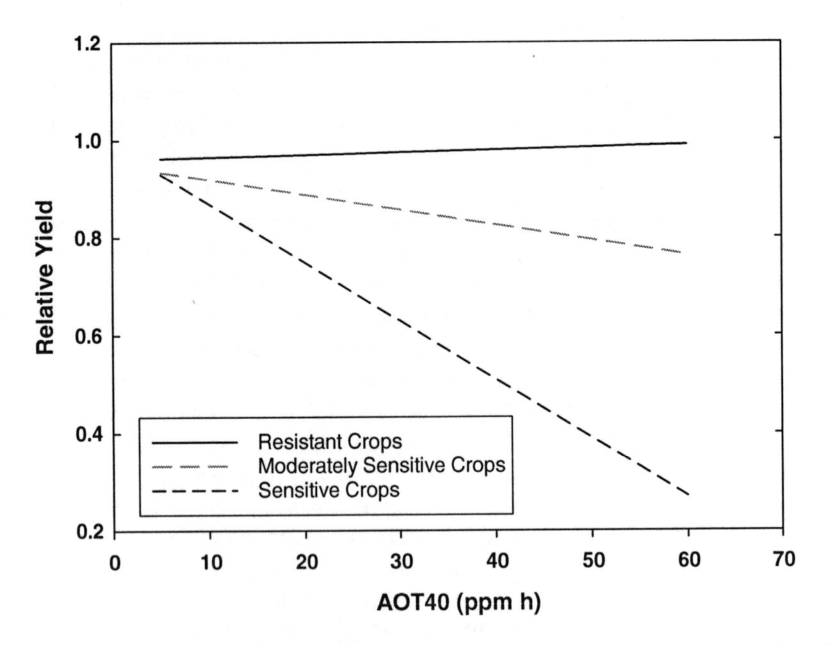

Fig. 7.4 The combined response of O_3-resistant crops (barley, plum and strawberry), moderately O_3-sensitive crops (sugar beet, potato, oilseed rape, tobacco, rice, maize, grape and broccoli) and O_3-sensitive (wheat, water melon, pulses, cotton, turnip, tomato, onion, soybean and lettuce) to O_3 dosage, measured as the accumulation over a threshold of 40 ppb (AOT 40) (figure is redrawn with permission from Mills et al. (2007)

emissions and legislation, and they also have a number of other limitations. First, they are based on crop response functions derived for European and North American crops that were grown under well-fertilized and well-watered conditions (Van Dingenen et al. 2009). Second, they do not take into account the interaction of rising [O_3] with changes in temperature, atmospheric [CO_2] and the hydrological cycle, which would affect O_3 uptake into the leaves (Wang and Mauzerall 2004). In most studies where crops have been grown in elevated [CO_2] and elevated [O_3], yield loss is less than with [O_3] alone (Morgan et al. 2003; Fuhrer 2009). Still, with a current cost of crop losses to O_3 in the range of $14–26 billion (Van Dingenen et al. 2009), further research on understanding the mechanisms of response and breeding for tolerance is critical.

7.7 Knowledge Gaps and Future Challenges

Our understanding of how crop yields will respond to rising atmospheric [CO_2] and [O_3] has improved substantially with the tremendous amount of research over the past four decades. However, a number of knowledge gaps and research challenges remain. FACE experiments have been restricted to temperate locations, with a limited selection of germplasm, which significantly restricts extrapolation of the results to global crop production estimates. Furthermore, FACE experiments have not investigated the interactive effects of simultaneous changes in [CO_2], temperature, soil moisture and [O_3] (Long et al. 2006). While technologically difficult, these experiments are not impossible. Infrared heater arrays (Kimball et al. 2008), passive infrared night-time warming and rain exclusion systems (Mikkelsen et al. 2008), and open-air O_3 enrichment systems (Morgan et al. 2004; Karnosky et al. 2007) have been used to investigate interactive effects of [CO_2] and other climate change factors. Apart from waiting 50–100 years to test model outputs, these experimental approaches remain the only way to test and constrain model projections of future food supply.

Adapting crops to elevated [CO_2] remains a major challenge (Ainsworth et al. 2008a, b). Studies of wheat cultivars released throughout the twentieth century suggest that the sensitivity of yield to [CO_2] has declined in more recently released cultivars (Ziska et al. 2004; Manderscheid and Weigel 1997). The relative sensitivity of wheat grain yield with a doubling of [CO_2] concentration was strongly correlated with an increase in tiller production, leaf area, and subsequent panicle formation, and the ability to form new tillers was more limited in recent cultivars (Ziska et al. 2004). So, it seems that traditional breeding has not selected for [CO_2] responsiveness, in fact, the opposite has occurred. Furthermore, breeding has not inadvertently selected for O_3 tolerance (Fiscus et al. 2005). Thus, there is a need to understand the complex mechanisms of yield response to [CO_2] and to use the genetic diversity available to improve responsiveness (Ainsworth et al. 2008a). With predictions that drought, high temperature stress and O_3 pollution will increase throughout this century, causing damage to crop production and making the timing and application of nutrients, herbicides and pesticides more difficult (Porter and Semenov 2005; Tubiello et al. 2007b), maximizing crop response to elevated [CO_2] is even more important.

7.8 Summary

- Rising atmospheric [CO_2] has a direct effect on crop carbon uptake and water use, and these direct effects feed forward to alter economic yield.
- Major C_3 grain crops show an approximate 13% increase in seed yield at ~550 ppm, the [CO_2] expected for 2050. However, C_4 crops do not show a significant yield increase at elevated [CO_2] under conditions of adequate water supply.
- Growth at elevated [CO_2] decreases grain protein and mineral content, which has significant implications for animal and human nutrition.
- Current tropospheric [O_3] causes significant losses of potential crop yields ($14–26 billion), and it is likely that [O_3] will rise and be a greater problem in the future.
- Our ability to accurately model future food supply depends critically on understanding crop responses to elevated [CO_2], and the interaction with other climate change factors, including rising [O_3], temperature and drought stress.

References

Ainsworth EA (2008) Rice production in a changing climate: a meta-analysis of responses to elevated carbon dioxide and elevated ozone concentration. Glob Change Biol 14:1642–1650

Ainsworth EA, Long SP (2005) What have we learned from 15 years of free-air CO_2 enrichment (FACE)? A meta-analytic review of the responses of photosynthesis, canopy. New Phytol 165:351–371

Ainsworth EA, Rogers A (2007) The response of photosynthesis and stomatal conductance to rising [CO_2]: mechanisms and environmental interactions. Plant Cell Environ 30:258–270

Ainsworth EA, Davey PA, Bernacchi CJ, Dermody OC, Heaton EA, Moore DJ, Morgan PB, Naidu SL, Ra HSY, Zhu XG, Curtis PS, Long SP (2002) A meta-analysis of elevated [CO_2] effects on soybean (*Glycine max*) physiology, growth and yield. Glob Change Biol 8:695–709

Ainsworth EA, Rogers A, Nelson R, Long SP (2004) Testing the "source-sink" hypothesis of down-regulation of photosynthesis in elevated [CO_2] in the field with single gene substitutions in *Glycine max*. Agric For Meteorol 122:85–94

Ainsworth EA, Beier C, Calfapietra C, Ceulemans R, Durand-Tardif M, Farquhar GD, Godbold DL, Hendrey GR, Hickler T, Kaduk J, Karnosky DF, Kimball BA, Koerner C, Koornneef M, Lafarge T, Leakey ADB, Lewin KF, Long SP, Manderscheid R, McNeil DL, Mies TA, Miglietta F, Morgan JA, Nagy J, Norby RJ, Norton RM, Percy KE, Rogers A, Soussana JF, Stitt M, Weigel HJ, White JW (2008a) Next generation of elevated [CO_2] experiments with crops: a critical investment for feeding the future world. Plant Cell Environ 31:1317–1324

Ainsworth EA, Leakey ADB, Ort DR, Long SP (2008b) FACE-ing the facts: inconsistencies and interdependence among field, chamber and modeling studies of elevated [CO_2] impacts on crop yield and food supply. New Phytol 179:5–9

Ainsworth EA, Rogers A, Leakey ADB (2008c) Targets for crop biotechnology in a future high-CO_2 and high-O_3 world. Plant Physiol 147:13–19

Akimoto H (2003) Global air quality and pollution. Science 302:1716–1719

Allen LH Jr, Boote KJ, Jones JW, Jones PH, Valle RR, Acock B, Rogers HH, Dahlman RC (1987) Response of vegetation to rising carbon dioxide: photosynthesis, biomass, and seed yield of soybean. Global Biogeochem Cy 1:1–14

Amthor JS (2001) Effects of atmospheric CO_2 concentration on wheat yield: review of results from experiments using various approaches to control CO_2 concentration. Field Crop Res 73:1–34

Arp WJ (1991) Effects of source-sink relations on photosynthetic acclimation to elevated CO_2. Plant Cell Environ 14:869–875

Ashmore MR (2005) Assessing the future global impacts of ozone on vegetation. Plant Cell Environ 28:949–964

Asseng S, Jamieson PD, Kimball B, Pinter P, Sayre K, Bowden JW, Howden SM (2004) Simulated wheat growth affected by rising temperature, increased water deficit and elevated atmospheric CO_2. Field Crop Res 85:85–102

Ball JT, Woodrow IE, Berry JA (1987) A model predicting stomatal conductance and its contribution to the control of photosynthesis under different environmental conditions. In: Biggens J (ed) Progress in photosynthesis research. Martinus-Nijhoff Publishers, Dordrecht, the Netherlands

Bannayan M, Kobayashi K, Kim HY, Lieffering M, Okada M, Miura S (2005) Modeling the interactive effects of atmospheric CO_2 and N on rice growth and yield. Field Crop Res 93:237–251

Bernacchi CJ, Kimball BA, Quarles DR, Long SP, Ort DR (2007) Decreases in stomatal conductance of soybean under open-air elevation of [CO_2] are closely coupled with decreases in ecosystem evapotranspiration. Plant Physiol 143:134–144

Bindi M, Fibbi L, Frabotta A, Chiesi M, Selveggi G, Magliulo V (1999) Free air CO_2 enrichment of potato (*Solanum tuberosum* L.). In: CHIP (ed) Changing climate and potential impacts on potato yield and quality: CHIP project, final report; contract ENV4-CT97-0489. Commission of the European Union, Brussels, Belgium

Bindi M, Miglietta F, Vaccari F, Magliulo E, Giuntola A (2006) Growth and quality responses of potato to elevated [CO_2]. In: Nosberger J, Long SP, Norby RJ, Stitt M, Hendrey GR, Blum H (eds) Managed ecosystems and CO_2, vol 187. Springer, Germany

Black VJ, Black CR, Roberts JA, Stewart CA (2000) Impact of ozone on the reproductive development of plants. New Phytol 147:421–447

Bowes G (1991) Growth at elevated CO_2: photosynthetic responses mediated through Rubisco. Plant Cell Environ 14:795–806

Canadell JG, Le Quere C, Raupach MR, Field CB, Buitenhuis ET, Ciais P, Conway TJ, Gillett NP, Houghton RA, Marland G (2007) Contributions to accelerating atmospheric CO_2 growth from economic activity, carbon intensity, and efficiency of natural sinks. Proc Natl Acad Sci USA 104:18866–18870

Conley MM, Kimball BA, Brooks TJ, Pinter PJ, Hunsaker DJ, Wall GW, Adam NR, LaMorte RL, Matthias AD, Thompson TL, Leavitt SW, Ottman MJ, Cousins AB, Triggs JM (2001) CO_2 enrichment increases water-use efficiency in sorghum. New Phytol 151:407–412

Cure JD, Acock B (1986) Crop responses to carbon dioxide doubling – a literature survey. Agricul For Meterol 38:127–145

De Souza AP, Gaspar M, Da Silva EA, Ulian EC, Waclawovsky AJ, Nishiyama MY, Dos Santos RV, Teixeira MM, Souza GM, Buckeridge MS (2008) Elevated CO_2 increases photosynthesis, biomass and productivity, and modifies gene expression in sugarcane. Plant Cell Environ 31:1116–1127

Dentener F, Stevenson D, Ellingsen K, Van Noije T, Schultz M, Amann M, Atherton C, Bell N, Bergmann D, Bey I, Bouwman L, Butler T, Cofala J, Collins B, Drevet J, Doherty R, Eickhout B, Eskes H, Fiore A, Gauss M, Hauglustaine D, Horowitz L, Isaksen ISA, Josse B, Lawrence M, Krol M, Lamarque JF, Montanaro V, Müller JF, Peuch VH, Pitari G, Pyle J, Rast S,

Rodriguez J, Sanderson M, Savage NH, Shindell D, Strahan S, Szopa S, Sudo K, Van Dingenen R, Wild O, Zeng G (2006) The global atmospheric environment for the next generation. Environ Sci Technol 40:3586–3594

Drake BG, Gonzalez-Meler MA, Long SP (1997) More efficient plants: a consequence of rising atmospheric CO_2? Annu Rev Plant Physiol Plant Mol Biol 48:609–639

Ewert F, Rodriguez D, Jamieson P, Semenov MA, Mitchell RAC, Goudriaan J, Porter JR, Kimball BA, Pinter PJ, Manderscheid R, Weigel HJ, Fangmeier A, Fereres E, Villalobos F (2002) Effects of elevated CO_2 and drought on wheat: testing crop simulation models for different experimental and climatic conditions. Agric Ecosyst Environ 93:249–266

Fangmeier A, De Temmerman L, Mortensen L, Kemp K, Burke J, Mitchell R, van Oijen M, Weigel HJ (1999) Effects on nutrients and on grain quality in spring wheat crops grown under elevated CO_2 concentrations and stress conditions in the European, multiple-site experiment 'ESPACE-wheat'. Eur J Agron 10:215–229

Farquhar GD, Dubbe DR, Raschke K (1978) Gain of feedback loop involving carbon dioxide and stomata – theory and measurement. Plant Physiol 62:406–412

Farquhar GD, von Caemmerer S, Berry JA (1980) A biochemical model of photosynthetic CO_2 assimilation in leaves of C_3 species. Planta 149:78–90

Feng Z, Kobaysahi K, Ainsworth EA (2008) Impact of elevated ozone concentration on growth, physiology, and yield of wheat (*Triticum aestivum* L.): a meta-analysis. Glob Change Biol 14:2696–2708

Fischer RA (2007) Understanding the physiological basis of yield potential in wheat. J Agric Sci 145:99–113

Fiscus EL, Booker FL, Burkey KO (2005) Crop responses to ozone: uptake, modes of action, carbon assimilation and partitioning. Plant Cell Environ 28:997–1011

Fisher GEJ (2008) Micronutrients and animal nutrition and the link between the application of micronutrients to crops and animal health. Turk J Agric For 32:221–233

Franzaring J, Hogy P, Fangmeier A (2008) Effects of free-air CO_2 enrichment on the growth of summer oilseed rape (Brassica napus cv. Campino). Ag Ecosys Environ 128:127–134

Fuhrer J (2009) Ozone risk for crops and pastures in present and future climates. Naturwissenschaften 96:173–194

Fuhrer J, Skarby L, Ashmore MR (1997) Critical levels for ozone effects on vegetation in Europe. Environ Poll 97:91–106

Fuhrer J, Booker F (2003) Ecological issues related to ozone: agricultural issues. Environ Int 29:141–154

Grant RF, Wall GW, Kimball BA, Frumau KFA, Pinter PJ, Hunsaker DJ, Lamorte RL (1999) Crop water relations under different CO_2 and irrigation: testing of ecosys with the free air CO_2 enrichment (FACE) experiment. Agric Forest Meteorol 95:27–51

Grossman-Clarke S, Pinter EJ, Kartschall T, Kimball BA, Hunsaker DJ, Wall GW, Garcia RL, LaMorte RL (2001) Modelling a spring wheat crop under elevated CO_2 and drought. New Phytol 150:315–335

Heagle AS, Philbeck RB, Ferrell RE, Heck WW (1989) Design and performance of a large, field exposure chamber to measure effects of air-quality on plants. J Environ Qual 18:361–368

Heck WW, Taylor OC, Tingey DT (1987) The NCLAN economic assessment: approach, findings and implications. In: Heck WW, Taylor OC, Tingey D (eds) Assessment of crop losses from air pollutants. Elsevier Applied Science, London

Hendrey GR, Miglietta F (2006) FACE technology: past, present, and future. In: Nosberger J, Long SP, Norby RJ, Stitt M, Hendrey GR, Blum H (eds) Managed ecosystems and CO_2, vol 187. Springer, Germany

Hendrey GR, Lewin KF, Nagy J (1993) Free air carbon-dioxide enrichment – development, progress, results. Vegetatio 104:17–31

Higgins TJ, Chrispeels MJ (2003) Plants in human nutrition and animal feed. In: Chrispeels MJ, Sadava DE (eds) Plants, genes and crop biotechnology, 2nd edn. Jones and Bartlett, USA

Hunsaker DJ, Kimball BA, Pinter PJ, Wall GW, LaMorte RL, Adamsen FJ, Leavitt SW, Thompson TL, Matthias AD, Brooks TJ (2000) CO_2 enrichment and soil nitrogen effects on wheat evapotranspiration and water use efficiency. Agric Forest Meteorol 104:85–105

Jamieson PD, Berntsen J, Ewert F, Kimball BA, Olesen JE, Pinter PJ, Porter JR, Semenov MA (2000) Modelling CO_2 effects on wheat with varying nitrogen supplies. Agric Ecosyst Environ 82:27–37

Karnosky DF, Werner H, Holopainen T, Percy K, Oksanen T, Oksanen E, Heerdt C, Fabian P, Nagy J, Heilman W, Cox R, Nelson N, Matyssek R (2007) Free-air exposure systems to scale up ozone research to mature trees. Plant Biol 9:181–190

Kartschall T, Grossman S, Pinter PJ, Garcia RL, Kimball BA, Wall GW, Hunsaker DJ, LaMorte RL (1995) A simulation of phenology, growth, carbon dioxide exchange and yields under ambient atmosphere and free-air carbon dioxide enrichment (FACE) Maricopa, Arizona, for wheat. J Biogeogr 22:611–622

Kim HY, Lieffering M, Miura S, Kobayashi K, Okada M (2001) Growth and nitrogen uptake of CO_2-enriched rice under field conditions. New Phytol 150:223–229

Kim HY, Lieffering M, Kobayashi K, Okada M, Mitchell MW, Gumpertz M (2003) Effects of free-air CO_2 enrichment and nitrogen supply on the yield of temperate paddy rice crops. Field Crop Res 83:261–270

Kimball BA (1983) Carbon dioxide and agricultural yield – an assemblage and analysis of 430 prior observations. Agron J 75:779–788

Kimball BA (2006) The effects of free-air [CO_2] enrichment of cotton, wheat, and sorghum. In: Nosberger J, Long SP, Norby RJ, Stitt M, Hendrey GR, Blum H (eds) Managed ecosystems and CO_2, vol 187. Springer, Germany

Kimball BA, Kobayashi K, Bindi M (2002) Responses of agricultural crops to free-air CO_2 enrichment. Adv Agron 77:293–368

Kimball BA, Conley MM, Wang S, Lin X, Luo C, Morgan J, Smith D (2008) Infrared heater arrays for warming ecosystem field plots. Glob Change Biol 14:309–320

Kobayashi K, Okada M, Kim HY, Lieffering M, Miura S, Hasegawa T (2006) Paddy rice responses to free-air [CO_2] enrichment. In: Nosberger J, Long SP, Norby RJ, Stitt M, Hendrey GR, Blum H (eds) Managed ecosystems and CO_2, vol 187. Springer, Germany

Leadley PW, Drake BG (1993) Open top chambers for exposing plant canopies to elevated CO_2 concentration and for measuring net gas-exchange. Vegetatio 104:3–15

Leakey ADB (2009) Rising atmospheric carbon dioxide concentration and the future of C_4 crops for food and fuel. Proc Royal Soc B 276:2333–2343

Leakey ADB, Bernacchi CJ, Dohleman FG, Ort DR, Long SP (2004) Will photosynthesis of maize (*Zea mays*) in the US Corn Belt increase in future [CO_2] rich atmospheres? An analysis of diurnal courses of CO_2 uptake under free-air concentration enrichment (FACE). Glob Change Biol 10:951–962

Leakey ADB, Bernacchi CJ, Ort DR, Long SP (2006) Long-term growth of soybean at elevated [CO_2] does not cause acclimation of stomatal conductance under fully open-air conditions. Plant Cell Environ 29:1794–1800

Leakey ADB, Ainsworth EA, Bernacchi CJ, Rogers A, Long SP, Ort DR (2009a) Elevated CO_2 effects on plant carbon, nitrogen and water relations: six important lessons from FACE. J Exp Bot, in press. doi:10.1093/jxb/erp096

Leakey ADB, Ainsworth EA, Bernard SM, Markelz RJC, Ort DR, Placella SA, Rogers A, Smith MD, Sudderth EA, Weston DJ, Wullschleger SD, Yuan S (2009b) Gene expression profiling – opening the block box of plant ecosystem responses to global change. Global Change Biol 15:1201–1213

Leakey ADB, Xu F, Gillespie KM, McGrath JM, Ainsworth EA, Ort DR (2009c) The genomic basis for stimulated respiration by plants growing under elevated carbon dioxide. Proc Natl Acad Sci USA 106:3597–3602

Lewin KF, Hendrey GR, Kolber Z (1992) Brookhaven national laboratory free-air carbon-dioxide enrichment facility. Crit Rev Plant Sci 11:135–141

Loladze I (2002) Rising atmospheric CO_2 and human nutrition: toward globally imbalanced plant stoichiometry? Trends Ecol Evol 17:457–461

Long SP (1999) Understanding the impacts of rising CO_2: the contribution of environmental physiology. In: Press MC, Scholes JD, Barker MG (eds) Physiological Plant Ecology. Blackwell Science, Oxford, UK

Long SP, Ainsworth EA, Rogers A, Ort DR (2004) Rising atmospheric carbon dioxide: plants face the future. Annu Rev Plant Biol 55:591–628

Long SP, Ainsworth EA, Leakey ADB, Morgan PB (2005) Global food insecurity. Treatment of major food crops with elevated carbon dioxide or ozone under large-scale fully open-air conditions suggests recent models may have overestimated future yields. Philos Trans R Soc B-Biol Sci 360:2011–2020

Long SP, Ainsworth EA, Leakey ADB, Nosberger J, Ort DR (2006) Food for thought: lower-than-expected crop yield stimulation with rising CO_2 concentrations. Science 312:1918–1921

Manderscheid R, Weigel HJ (1997) Photosynthetic and growth responses of old and modern spring wheat cultivars to atmospheric CO_2 enrichment. Agric Ecosyst Environ 64:65–73

Manderscheid R, Bender J, Jager HJ, Weigel HJ (1995) Effects of season long CO_2 enrichment on cereals. 2. Nutrient concentrations and grain quality. Agric Ecosyst Environ 54:175–185

Mauzerall DL, Wang X (2001) Protecting agricultural crops from the effects of tropospheric ozone exposure: reconciling science and standard setting in the United States, Europe, and Asia. Annu Rev Energy Environ 26:237–268

McConnaughay KDM, Berntson GM, Bazzaz FA (1993) Limitations to CO_2-induced growth enhancement in pot studies. Oecologia 94:550–557

McLeod AR, Long SP (1999) Free-air carbon dioxide enrichment (FACE) in global change research: a review. Adv Ecol Res 28:1–56

Meehl GA, Stocker TF, Collins WD et al. (2007) Global climate projections. In: Solomon S, Qin D, Manning M et al. (eds) Climate Change 2007: The Physical Science Basis. Contribution of Working Group I to the Fourth Assessment Report of the Inter-Governmental Panel on Climate Change. Cambridge University Press, Cambridge, UK and New York, NY, USA

Miglietta F, Magliulo V, Bindi M, Cerio L, Vaccari FP, Loduca V, Peressotti A (1998) Free air CO_2 enrichment of potato (*Solanum tuberosum* L.): development, growth and yield. Glob Change Biol 4:163–172

Miglietta F, Peressotti A, Vaccari FP, Zaldei A, deAngelis P, Scarascia-Mugnozza G (2001) Free-air CO_2 enrichment (FACE) of a poplar plantation: the POPFACE fumigation system. New Phytol 150:465–476

Mikkelsen TN, Beier C, Jonasson S, Holmstrup M, Schmidt IK, Ambus P, Pilegaard K, Michelsen A, Albert K, Andresen LC, Arndal MF, Bruun N, Christensen S, Danbaek S, Gundersen P, Jorgensen P, Linden LG, Kongstad J, Maraldo K, Prieme A, Riis-Nielsen T, Ro-Poulsen H, Stevnbak K, Selsted MB, Sorensen P, Larsen KS, Carter MS, Ibrom A, Martinussen T, Miglietta F, Sverdrup H (2008) Experimental design of multifactor climate change experiments with elevated CO_2, warming and drought: the CLIMAITE project. Funct Ecol 22:185–195

Mills G, Buse A, Gimeno B, Bermejo V, Holland M, Emberson L, Pleijel H (2007) A synthesis of AOT40-based response functions and critical levels of ozone for agricultural and horticultural crops. Atmos Environ 41:2630–2643

Morgan PB, Ainsworth EA, Long SP (2003) How does elevated ozone impact soybean? A meta-analysis of photosynthesis, growth and yield. Plant Cell Environ 26:1317–1328

Morgan PB, Bernacchi CJ, Ort DR, Long SP (2004) An in vivo analysis of the effect of season-long open-air elevation of ozone to anticipated 2050 levels on photosynthesis in soybean. Plant Physiol 135:2348–2357

Morgan PB, Bollero GA, Nelson RL, Dohleman FG, Long SP (2005) Smaller than predicted increase in aboveground net primary production and yield of field-grown soybean under fully open-air [CO_2] elevation. Glob Change Biol 11:1856–1865

Morrison MJ, Voldeng HG, Cober ER (2000) Agronomic changes from 58 years of genetic improvement of short-season soybean cultivars in Canada. Agron J 92:780–784

Nowak RS, Ellsworth DS, Smith SD (2004) Functional responses of plants to elevated atmospheric CO_2 - do photosynthetic and productivity data from FACE experiments support early predictions? New Phytol 162:253–280

Ottman MJ, Kimball BA, Pinter PJ, Wall GW, Vanderlip RL, Leavitt SW, LaMorte RL, Matthias AD, Brooks TJ (2001) Elevated CO_2 increases sorghum biomass under drought conditions. New Phytol 150:261–273

Pang J, Zhu JG, Xie ZB, Liu G, Zhang YL, Chen GP, Zeng Q, Cheng L (2006) A new explanation of the N concentration decrease in tissues of rice (*Oryza sativa* L.) exposed to elevated atmospheric pCO_2. Environ Exp Bot 57, 98–105

Parry M, Rosenzweig C, Iglesias A, Fischer G, Livermore M (1999) Climate change and world food security: a new assessment. Glob Environ Change 9:S51–S67

Parry ML, Rosenzweig C, Iglesias A, Livermore M, Fischer G (2004) Effects of climate change on global food production under SRES emissions and socio-economic scenarios. Glob Environ Change 14:53–67

Parry M, Rosenzweig C, Livermore M (2005) Climate change, and risk global food supply of hunger. Phil Trans R Soc B 360:2125–2138

Peart RM, Jones JW, Curry RB, Boote K, Allen LH (1989) Impact of climate change on crop yield in the southeastern USA: a simulation study. In: Smith JB, Tirpak DA (eds.) The Potential Effects of Global Climate Change on the United States, vol. 1. EPA, Washington, DC

Pinter PJ, Kimball BA, Garcia RL, Wall GW, Hunsaker DJ, LaMorte RL (1996) Free air CO_2 enrichment: responses of cotton and wheat crops. In: Koch GW, Mooney HA (eds) Carbon dioxide and terrestrial ecosystems. Academic, San Diego

Porter JR, Semenov MA (2005) Crop responses to climatic variation. Phil Trans R Soc B 360:2021–2035

Prather M, Gauss M, Berntsen T, Isaksen I, Sundet J, Bey I, Brasseur G, Dentener F, Derwent R, Stevenson D, Grenfell L, Hauglustaine D, Horowitz L, Jacob D, Mickley L, Lawrence M, von Kuhlmann R, Muller J-F, Pitari G, Rogers H, Johnson M, Pyle J, Law K, van Weele M, Wild O (2003) Fresh air in the 21st century. Geophys Res Lett 30:1100

Raupach MR, Marland G, Ciais P, Le Quere C, Canadell JG, Klepper G, Field CB (2007) Global and regional drivers of accelerating CO_2 emissions. Proc Natl Acad Sci USA 104:10288–10293

Rogers HH, Bingham GE, Cure JD, Smith JM, Surano KA (1983) Responses of selected plant species to elevated carbon dioxide in the field. J Env Qual 12:569–574

Sasaki H, Hara T, Ito S, Miura S, Hoque MM, Lieffering M, Kim HY, Okada M, Kobayashi K (2005) Seasonal changes in canopy photosynthesis and respiration, and partitioning of photosynthate, in rice (*Oryza sativa* L.) grown under free-air CO_2 enrichment. Plant Cell Physiol 46:1704–1712

Seneweera SP, Conroy JP (1997) Growth, grain yield and quality of rice (*Oryza sativa* L.) in response to elevated CO_2 and phosphorus nutrition (Reprinted from Plant nutrition for sustainable food production and environment, 1997). Soil Sci Plant Nutr 43:1131–1136

Seneweera SP, Conroy JP, Ishimaru K, Ghannoum O, Okada M, Lieffering M, Kim HY, Kobayashi K (2002) Changes in source-sink relations during development influence photosynthetic acclimation of rice to free air CO_2 enrichment (FACE). Funct Plant Biol 29:945–953

Shimono H, Okada M, Yamakawa Y, Nakamura H, Kobayashi K, Hasegawa T (2007) Lodging in rice can be alleviated by atmospheric CO_2 enrichment. Agric Ecosyst Environ 118:223–230

Siegenthaler U, Stocker TF, Monnin E, Luthi D, Schwander J, Stauffer B, Raynaud D, Barnola JM, Fischer H, Masson-Delmotte V, Jouzel J (2005) Stable carbon cycle-climate relationship during the late Pleistocene. Science 310:1313–1317

Solomon S, Qin D, Manning M, Alley RB, Berntsen R, Bindoff NL, Chen Z, Chidthaisong A, Gregory JM, Hegerl GC, Heimann M, Hewitson B, Hoskins BJ, Joos F, Jouzel J, Kattsov V, Lohmann U, Matsuno T, Molina M, Nicholls N, Overpeck J, Raga G, Ramaswamy V, Ren J, Rusticucci M, Somerville R, Stocker TF, Whetton P, Wood RA, Wratt D (2007) Technical Summary. In: Solomon S, Qin D, Manning M, Chen Z, Marquis M, Averyt KB, Tignor M, Miller HL (eds) Climate change 2007: the physical science basis. Contribution of working group I to the fourth assessment report of the intergovernmental panel on climate change. Cambridge University Press, Cambridge, UK and New York, NY, USA

Taub DR, Miller B, Allen H (2008) Effects of elevated CO_2 on the protein concentration of food crops: a meta-analysis. Glob Change Biol 14:565–575

Tong D, Mathur R, Schere K, Kang D, Yu S (2007) The use of air quality forecasts to assess impacts of air pollution on crops: methodology and case study. Atmos Environ 41:8772–8784

Tubiello FN, Ewert F (2002) Simulating the effects of elevated CO_2 on crops: approaches and applications for climate change. Eur J Agron 18:57–74

Tubiello FN, Rosenzweig C, Kimball BA, Pinter PJ, Wall GW, Hunsaker DJ, LaMorte RL, Garcia RL (1999) Testing CERES-wheat with free-air carbon dioxide enrichment (FACE) experiment data: CO_2 and water interactions. Agron J 91:247–255

Tubiello FN, Amthor JS, Boote KJ, Donatelli M, Easterling W, Fischer G, Gifford RM, Howden M, Reilly J, Rosenzweig C (2007a) Crop response to elevated CO_2 and world food supply: a comment on "Food for Thought." by Long et al., Science 312: 1918–1921, 2006. Eur J Agron 26:215–223

Tubiello FN, Soussana JF, Howden SM (2007b) Crop and pasture response to climate change. Proc Natl Acad Sci USA 104:19686–19690

United Nations, Administrative Committee on Co-ordination, Sub-committee on N, International Food Policy Research I (2000) 4th report on the world nutrition situation: nutrition throughout the life cycle. United Nations, Administrative Committee on Coordination, Subcommittee on Nutrition (ACC/SCN). International Food Policy Research Institute (IFPRI), Geneva, Switzerland; Washington, DC

Van Dingenen R, Dentener FJ, Raes F, Krol MC, Emberson L, Cofala J (2009) The global impact of ozone on agricultural crop yields under current and future air quality legislation. Atmos Env 43:604–618

Vingarzan R (2004) A review of surface ozone background levels and trends. Atmos Environ 38:3431–3442

Volz A, Kley D (1988) Evaluation of the Montsouris series of ozone measurements made in the 19th century. Nature 332:240–242

von Caemmerer S, Furbank RT (2003) The C_4 pathway: an efficient CO_2 pump. Photosynth Res 77:191–207

Wang X, Mauzerall DL (2004) Characterizing distributions of surface ozone and its impact on grain production in China, Japan and South Korea: 1990 and 2020. Atmos Environ 38:4383–4402

Weigel HJ, Pacholski A, Waloszczyk K, Fruhauf C, Manderscheid R, Anderson TH, Heinemeyer O, Kleikamp B, Helal M, Burkart S, Schrader S, Sticht C, Giesemann A (2006) Effects of elevated atmospheric CO_2 concentrations on barley, sugar beet and wheat in a rotation: examples from the Braunschweig carbon project. Landbauforschung Volkenrode 56:101–115

Whitehead D, Hogan KP, Rogers GND, Byers JN, Hunt JE, McSeveny TM, Hollinger DY, Dungan RJ, Earl WB, Bourke MP (1995) Performance of large open-top chambers for long-term field investigations of tree response to elevated carbon dioxide concentration. J Biogeogr 22:307–313

Yang LX, Huang JY, Yang HJ, Dong GC, Liu G, Zhu JG, Wang YL (2006) Seasonal changes in the effects of free-air CO_2 enrichment (FACE) on dry matter production and distribution of rice (*Oryza sativa* L.). Field Crop Res 98:12–19

Ziska LH, Morris CF, Goins EW (2004) Quantitative and qualitative evaluation of selected wheat varieties released since 1903 to increasing atmospheric carbon dioxide: can yield sensitivity to carbon dioxide be a factor in wheat performance? Glob Change Biol 10:1810–1819

Part III

Chapter 8
Food Security and Adaptation to Climate Change: What Do We Know?

Marshall Burke and David Lobell

Abstract The potential for agricultural systems to adapt to climate change is at once both promising and poorly understood. This chapter reviews possible producer and consumer responses to a changing climate, the ability of these responses to offset otherwise negative impacts on food security, and the role of public and private institutions in investing in adaptation where individual responses are insufficient. Accumulated evidence suggests that wealthier societies and households will be better able to adapt to a changing climate because of their greater availability of alternatives and their ability to take advantage of them. Accordingly, investments that improve options for the poor, such as improved agricultural production technologies, financial instruments, and off-farm income opportunities, will likely be critical for adapting food security to a changing climate.

8.1 Introduction

Climate change will not confront a static world. Humans respond to changes in their natural and economic environment and often make themselves better off by doing so, a responsiveness clearly evident in agriculture. As human populations grew and spread over past millennia, food production was expanded into far corners of the world, feeding growing populations in strikingly diverse environments and climates. This ability of humanity to adapt agriculture to new climates is evidence to many that climate change poses no fundamental threat to agriculture – that clever humans, as in centuries past, will simply adapt agriculture to its new growing conditions.

But the magnitude and speed of climate change that is expected over the next century raises serious questions about how much agriculture can be adapted to new climates, how quickly, and at what cost. Will simple farm-level measures such as switching crop varieties be enough to offset expected losses in much of the world? Or will larger investments in crop breeding or irrigation infrastructure be needed to meet

M. Burke (✉) and D. Lobell
Stanford University, CA, USA

D. Lobell and M. Burke (eds.), *Climate Change and Food Security*, Advances in Global Change Research 37, DOI 10.1007/978-90-481-2953-9_8, the food needs of a growing global population? Or could even these efforts fall short? Such questions are central to both anticipating the full impacts of climate change on food security and human livelihoods, and in planning appropriate responses.

This chapter will explore potential adaptations to climate change that might improve food security, where "adaptation" is understood to mean any response that improves an outcome (Reilly and Schimmelpfennig 2000). Many possible adaptations involve direct changes to agricultural systems, such as changing when and where crops are grown. But because food security involves much more than just food production (Chapter 2), we also consider various broader responses to climate change that might improve food security, such as improving social safety nets that protect the poor in adverse years.

Of central interest is the potential of these measures to offset many of the anticipated negative effects of climate change on food security, and in particular the extent to which such adaptations will happen more or less on their own (so called 'autonomous adaptation') as opposed to requiring significant investment and foresight for them to occur ('planned adaptation'). For instance, if we expect farmers to automatically recognize climate shifts and react in ways that offset expected losses, then the need for outside investment and policy intervention in adaptation is small. But if we expect farmers to have trouble responding on their own, and that this inability appears to threaten global or regional food security, then there would seem a pressing need to understand what broader investments in adaptation would be required.

Unfortunately, there is little existing quantitative evidence on the ability of adaptation to improve food security outcomes in the face of climate change, with large uncertainties surrounding both the potential gains from various adaptation measures and the extent to which they will be undertaken autonomously. Particularly difficult is disentangling the relationship between farmer responses to climate variability, which occur continually, and their likely longer run responses to changes in mean climate. Below we review the existing theory and evidence surrounding agricultural adaptation to climate change, and attempt to draw lessons both for investment priorities and for future research needs.

8.2 Farmer Adaptation to Climate: Dealing with Variability

The explicit focus of this book is on climate change – i.e. the potential shifts in the longer-run mean and extremes of temperature, precipitation, and other meteorological variables in a given area. And while longer-run climate exerts significant influence on agricultural decision-making, affecting what crops farmers grow and when and where they grow them, the actual amount of food produced in a given year depends on the specific realization of meteorological variables in that year. Year-to-year changes in these variables (or "climate variability") play a central role in global and regional food systems and in food security outcomes.

As a result, climate variability can both illuminate and constrain possible longer-run adaptation to climate change. For instance, farmer and food system responses to past weather events are some of the only evidence we have to understand how farmers respond to climate shifts. At the same time, variability also makes production more risky, which might inhibit risk averse farmers from undertaking broader adaptation measures. Finally, the year-to-year noise of climate variability might make it harder to recognize that climate is actually changing.

Observed farmer adaptations to climate variability fall into two main camps: ex ante measures, for which action is taken in anticipation of a given climate realization, and ex post responses, which are undertaken after the event is realized. Ex ante adaptations to variability often center around strategies of diversification, which attempt to capitalize on the differential effects that a given climate event might have on different crops and activities in a given year (Pandey et al. 2007). For instance, farmers growing rainfed crops in a drought-prone environment might seek to diversify the location of their farm plots to take advantage of the high spatial variability of rainfall, grow a range of crops or crop varieties with different sensitivities to climate, or to diversify income sources into non-farm enterprises that are less sensitive to climate (Pandey et al. 2007). They could also choose to maintain flexibility with regard to input decisions until uncertainties about weather realizations are reduced, for instance by shifting when crops are planted. Where possible, farmers might also pay to insure their harvests against failure.

Farmers also undertake various ex post strategies to decrease crop or welfare losses once climate events have been realized. Such strategies include drawing down cash reserves or stores of grain, borrowing from formal or informal credit markets or family, selling assets such as livestock, or migrating elsewhere in search for work in non-affected regions. Ex post adaptations can also include changes to management after the growing season has started, such as replanting of faster-maturing varieties if early-season planting fails, or irrigating where possible if rainfall is meager.

Not all strategies are available to all farmers unfortunately, nor are the available strategies always successful in buffering food security against a variable climate. In wealthier countries, farmers rarely go hungry as a result of drought or other adverse climate events. The existence of social safety nets and functioning financial markets ensure that farmers are either insured against losses, can borrow around them, or can receive help from the government to maintain livelihoods during bad times. Similarly, consumers in rich countries spend only a small percentage of their income on food, and are thus not very sensitive to the food price increases that often accompany droughts or floods.

The same is not often true in poor countries. Although both ex-post and ex-ante strategies can reduce climate-associated losses to some degree, the poorest households in particular are often unable to fully shield consumption from the effects of climate variability. This inability can be dramatic and devastating, as in the case of the drought-related famines in the Sahel and Horn of Africa in the 1980s, but they can also be more subtle, such as in the longer run documented negative effects of climate variability on health and economic outcomes in agricultural households, particularly for women and children (Hoddinott and Kinsey 2001; Maccini and Yang 2008). Such effects are realized because ex ante measures are insufficient, or ex post measures such as insurance or savings are unavailable, or both.

Also important are the perverse longer run effects of some of these adaptive measures on the food security of poor households. For instance, while ex ante strategies can reduce the risk of catastrophic losses in bad years, they can also reduce the income earned in good years, because farmers might have planted a less-risky but lower-yielding (and typically lower-value) crop. The long-run costs in foregone income from this risk-mitigation can be high – as much as 15–30% of average income (Rosenzweig and Binswanger 1993; Dercon and R. World Institute for Development Economics, 2002). Similarly, ex-post strategies can also avoid devastating declines in consumption in ways that harm longer run earning potential. Distressed liquidation of productive assets such as livestock or land can prop up consumption in one year, but dampen the subsequent productivity and food access of households in later years, an effect again well documented in the developing world. These perverse temporal tradeoffs are a perennial and painful dilemma faced by farmers throughout much of the developing world.

Given the negative impacts of climate variability on economic livelihoods and food security in much of the developing world, helping farmers better adapt to this variability is a central concern of development. Many have also argued that a focus on adapting to climate variability is the best way to approach adapting to climate change. This is in part because most farmers and governments can more readily understand the threat of variability, and thus are more likely to engage in building knowledge and institutional capacity to cope with variability (Washington et al. 2006; Cooper et al. 2008). It is also because climate variability can have large effects on livelihoods, and thus that longer-run adaptations will only be undertaken if they do not compromise the ability to cope with variability.

But as climate change adds to the stress of variability, will existing coping mechanisms be enough to offset expected losses from climate change in the absence of adaptation? Are current strategies for adapting to variability appropriate strategies for adapting to longer-run climate change? If not, and novel adaptations are called for, should we expect farmers to adopt them on their own, or will significant investment and policy intervention be needed to adapt food production to new climates?

8.3 Adapting to Climate Change: Some Difficulties

8.3.1 Signal Detection

Adaptation at the farmer level requires three basic steps: detecting a shift in one's external environment, determining that it would favor a change in behavior, and undertaking that change (Hanemann 2000; Kandlikar and Risbey 2000). Thus the first step in adapting to climate change requires detecting the signal of climate change in the noise of climate variability. Given the amplitude of climate variability in many regions, this might be no small task.

Figure 8.1 illustrates this detection problem, showing historical and projected future trends in temperature and precipitation for millet areas in Niger based on the GFDL climate model, which happens to project much larger decreases in precipita-

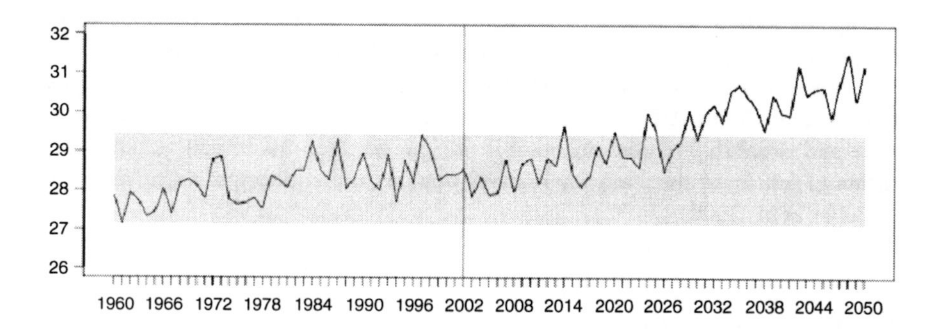

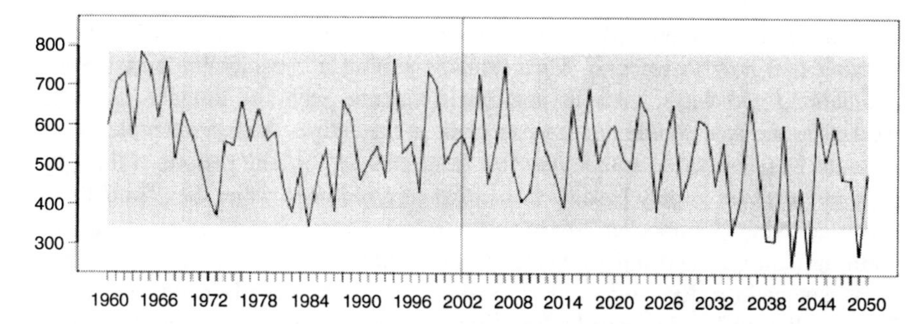

Fig. 8.1 Historical and projected future changes in temperature (*top panel*, in °C) and precipitation (*bottom panel*, in mm) for millet growing areas in Niger, 1960–2050. Data left of the vertical line are observed (CRU), and data to the right are based on projected changes from the GFDL climate model for the A1B scenario, assuming similar variability to the historical data. *Grey boxes* represent the range of historical variability between 1960 and 2002

tion in the Sahel (around 25% declines by 2050) than most other climate models. For temperature (top panel), the signal of climate change quickly emerges from the noise of past temperature variability, with every growing season hotter than the hottest year on record after around 2030 – a result we should expect for much of the tropics (Battisti and Naylor 2009). This is not the case with precipitation. Despite a very large projected decrease in average annual precipitation for millet-growing regions in Niger in this model, most years remain well within historical variability, potentially obscuring the underlying drying trend.

Farmers in developed countries have access to a wealth of climate and weather data, and so presumably could learn about trends in climate without having to sense them independently. The same is often not true for farmers in poorer countries, who rely on various traditional methods for climate forecasting, and who might be more or less on their own in discerning longer-run climate shifts.

Evidence is mixed on farmers' ability to correctly perceive such longer-run shifts. Meze-Hausken (2004) finds that farmers in northern Ethiopia report a decline in rainfall where rainfall gauges report no change. Maddison (2007) shows mixed results in farmers' ability to correctly perceive climate shifts across a range of African countries, with farmers in many countries correctly recognizing trends in mean temperature and rainfall, and others reporting trends in disagreement with observed climate data. Thomas et al. (2007) find qualitative evidence of South African farmers' abilities to detect subtle changes in mean state and variability of climate, but it is unclear whether this reveals actual recognition of trends, or the tendency to overestimate the frequency of negative events (Cooper et al. 2008).

8.3.2 Cognitive Biases

Once a farmer is convinced that the climate has changed, he or she must decide whether and how to respond. Most humans exhibit a considerable bias towards maintaining old ways, even in new environments, with the thought that what worked in the past should continue to work in the future. A clear example of this from the business world is that very few firms survive for long periods of time; the economy evolves largely by new firms replacing old ones rather than firms themselves adapting (Beinhocker 2006).

In agriculture, there may be a tendency to underestimate the need to change management in a new climate. For example, a survey recently conducted in the Yaqui Valley of Mexico asked wheat farmers whether they perceived a change in temperatures over the last decade, whether this change was positive or negative, and whether it had a positive, negative, or neutral effect on their yields (Ortiz-Monasterio and Lobell, 2005). Out of 88 farmers, 85 (or 97%) reported a significant shift in temperature, but only 33 (or 38%) felt the change had an effect on wheat yields, despite the fact that temperatures exert a strong control on yields in this region (Lobell et al. 2005).

Other surveys suggest an opposite problem: that farmers might be too quick to update their beliefs about changes in climate. In surveys of Canadian corn farmers, Smit et al. (1997) show that these farmers tend to heavily weight the previous year's weather in deciding what varieties to plant for the upcoming season. Though surveys are an imperfect means to gauging farmers' perceptions, these results illustrate that recognition of a climate trend is only one step towards successful adaptation.

8.4 Farmer Adaptations and Their Potential Gains

Supposing for now that a climate signal is detected, and that the need for a change in management is perceived, farmers must then decide how to respond. This response will depend on the choices they see themselves having and the perceived costs and benefits associated with each choice. Various potential adaptations are listed in Table 8.1, each of which we now explore in turn.

Table 8.1 Potential farmer adaptations to climate, and some reasons why they might or might not help

Adaptation	Why it might help	Why it might not help
Shift planting date	Take advantage of lengthened growing season	Less useful where current growing season length is not limited by cold temperatures
Switch varieties	Other existing varieties better suited to new climates	More suitable varieties not always available
Switch crops	Other crops more suitable to new climates	Hot countries have nothing to switch to
Expand area	Climate change could expand suitable area	Less true in the tropics; possible soil constraints; expansion may come with significant environmental costs
Expand irrigation	Helps alleviate moisture constraints	Can be expensive; often requires large government investment; many places have limited water resources
Diversify income	Non-farm income sources less climate sensitive	Rural non-farm economy linked to agricultural productivity
Migrate	Some areas might be hurt less than others by climate change	Urban areas already strained

8.4.1 Switching Planting Date

Perhaps the simplest farmer adaptations have to do with changes in on-farm management, which include decisions about what crops to grow and when and how to grow them. One of the more straightforward of these possible adaptations is the option to shift when in the year crops are planted. Current decisions about when to plant are made based on a number of factors, including available soil moisture, the expected timing of temperature extremes, and the demands of multi-cropped systems. Year-to-year shifts in planting dates are already a demonstrated farmer adaptation in the face of climate variability, particularly for farmers in rainfed environments who often must wait for the onset of the rainy season in order to plant. Farmers in parts of Africa and Asia, for instance, routinely shift planting dates by a month or more from year to year in response to variability in when monsoon rains arrive (Falcon et al. 2004; Tadross et al. 2005).

If climate change results in large shifts in the factors that determine optimal planting times, farmers could potentially gain by further changing the timing of their crop production. In a crop model simulation of US rainfed spring wheat under a warmer and wetter future climate, Tubiello et al. (2002) find that systematically shifting planting 2 weeks earlier transforms what would have been 20–25% yield losses by 2030 into modest gains. This is because cold temperatures limit early planting in current climate, subjecting the crop to heat and drought stress during critical stages of plant growth, and warmer climates appear to allow earlier planting and less stress during sensitive growth stages. Similarly, cropping systems where

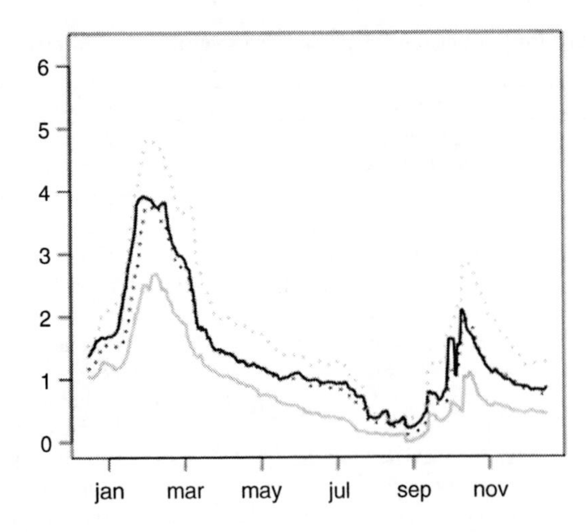

Fig. 8.2 Simulated maize yields (t/ha) in southeastern Kenya using CERES-maize. Planting is simulated independently on each day of the year, for current and hypothetical future climates. *Black solid line* = current climate; *black dotted*=+2°C; *grey solid*=+2°C, –20% precipitation; *grey dotted*=+2°C, +20% precipitation. Optimal planting for all scenarios peaks near the start of the long rains

irrigation is possible for much of the year might also benefit from shifting planting dates, particularly for crops likely to experience frequent temperature extremes in their current growing season as the climate warms.

But for rainfed systems throughout much of the tropics, where planting is typically limited by moisture rather than temperature, it is less clear that shifts in planting date will offset much of the expected damages from climate change – largely because climate change is expected to reduce growing season length throughout much of the tropics (Chapter 3). Figure 8.2 shows representative results for maize at a somewhat arid site in southeastern Kenya, with yields simulated using CERES-Maize for every possible planting date in the year under current and hypothetical future climates. The planting dates resulting in maximum yield occur near the beginning of the long rains, as expected, with a second smaller peak during the short rains (when a second crop is often planted). With planting moisture-limited, future climates suggest gains or losses in yield but no shifts in optimal planting date.

8.4.2 Switching Varieties or Crops

A second possible farmer adaptation to climate change is to switch varieties or crops to something better suited to the new climates they face. A farmer currently growing maize might switch to a faster-maturing maize variety if drought becomes more common, or might choose to grow a potentially more drought-tolerant crop like sorghum. But such decisions will not be made on the basis of climate alone. Different varieties and crops have different input requirements and costs associated with their production, different responsiveness to local stressors and can face very different output prices in ways that affect their profitability. To the extent that climate change affects the relative profitability of different crops and varieties in ways apparent to farmers – and in ways they can respond easily to – crop or variety switching could constitute a fruitful adaptation strategy.

In the case of switching varieties, climate change suggests two primary adaptation alternatives, the choice of which depends on whether moisture or heat is expected to be limiting. In low-rainfall areas where moisture stress is expected to remain a primary constraint on plant growth, a promising adaptation might be to plant faster-maturing varieties that avoid drought or heat stress during sensitive stages of plant growth, such as flowering or grain filling. Developing faster-maturing varieties for areas with short and variably rainy seasons (i.e. much of Africa) is a common goal of many breeding programs, and such a strategy would seem promising anywhere climate change is expected to shorten growing seasons.

In areas where moisture regimes exhibit little change, however, a move in the opposite direction toward longer maturing varieties might be preferred, because warmer temperatures tend to speed development and lower yields (Chapter 4). Longer maturing varieties would thus be required to maintain the length of time for total crop development as temperatures warm. Simulation studies indicate some benefits for this strategy. For instance, Tubiello et al. (2002) find that switching to longer-maturing winter wheat varieties at a site with plentiful moisture fully offsets the 15% projected yield losses under climate change, but find somewhat smaller gains for more arid areas.

Beyond shifting among varieties, farmers could also switch what crops they grow as the climate changes. As with choice of variety, farmers' choices about what crops to grow depend only partly on climate, and year-to-year crop choice decisions are likely dictated much more by expected prices at harvest than by climate concerns. For instance, farmers in the Midwestern US readily shift area between maize and soybeans depending on market signals. Nevertheless, over the long run climate exerts clear influence on crop choice. Climate clearly explains much of why rice is grown in the warm wet climates of Southeast Asia and wheat in the cooler, drier northern temperate latitudes of North America and Europe, and not the reverse. Similarly, the highly variable climates of much of Africa induce poor risk-averse farmers to grow lower-value but drought-tolerant crops such as cassava.

If climate matters to crop choice, then farmers could plausibly gain by switching crops if new climates favor a different crop over the one currently grown. This is the basic thrust of the so-called "Ricardian" estimates of climate change impacts on agriculture (Chapter 6). Instead of determining the potential impacts of climate change on the yield of a specific crop, as many studies do, these studies seek to isolate the effect of mean climate on land values in a given region, while controlling for other factors beyond climate that might affect land value (slope, soil type, etc.). The argument is that with well functioning markets, the value of land should reflect the current and (discounted) future stream of profits that can be made from using the land – whether it be used to grow corn or wheat or golf courses. The estimated effect of climate on land values should then in theory reflect all of the crop-switching adaptations farmers could make over the long run (Mendelsohn et al. 1994).

Consistent with the argument that the land values approach offers more thorough picture of farmer adaptation, estimated impacts of climate change are often more positive/less negative in these studies than in other studies that focus on single crops (e.g., Cline 2007, Chapter 5). But this method of modeling adaptation is not without its significant critics, who point out among other things that such methods might overstate the choice set that each individual farmer might have (Hanemann 2000), and thus overstate potential gains from adaptation.

More broadly, there are various factors that might constrain a farmer's ability or willingness to switch varieties or crops, such as the limited availability of alternatives, or the costs or perceived risks associated with adopting a new crop or variety. For instance, seed systems throughout much of Africa are poorly developed, such that locally adapted varieties of different maturity lengths or resistance to various abiotic stresses are not always available – and where they are developed, poor, risk averse farmers are often slow to adopt new technologies. Further, farming systems and local consumer taste preferences are often strongly intertwined, likely inhibiting rapid switching among crops. Finally, in countries with recurrent droughts but where temperatures will warm significantly under climate change (i.e. most of Africa), the optimal variety choice might be far from apparent: choose a shorter-maturing variety that avoids big losses in very dry years, or a longer-maturing variety that might maintain average yields as the climate warms?

These constraints are typically not captured in simulation studies of farmer adaptation, such as those using crop models, but can be picked up in some statistical approaches (Chapter 6). The limited evidence available from these approaches suggests that even in rich countries the potential for farmer adaptation within crops could be limited. For instance, using county-level data on US rainfed corn yields, Fisher et al. (2007) show that the estimated effect of temperature on yields is nearly equivalent whether you look at short run yield responses to variability (where little adaptation would be possible) or responses of yield to longer-run climate averages (under which farmers would have had time to adapt). This suggests that, at least under the range of existing technology and management, switching corn varieties would do little to stem the harmful effects of rising temperatures (see Chapter 6).

8.4.3 Expanding Irrigated and Total Cropped Area

In addition to changing their crop mix, farmers could also change how much land they farm or the way in which they farm what they have. Introducing irrigation into currently rainfed systems is an often cited adaptation option, and will indeed likely be critical for some regions. As mentioned, irrigation not only alleviates water stress but could expand the opportunities for switching planting dates and varieties, as well as increasing returns on investments in fertilizer and other inputs. Large scale expansions of irrigation infrastructure are typically financed and regulated by the public sector, and therefore farmers often cannot decide on their own to implement irrigation.

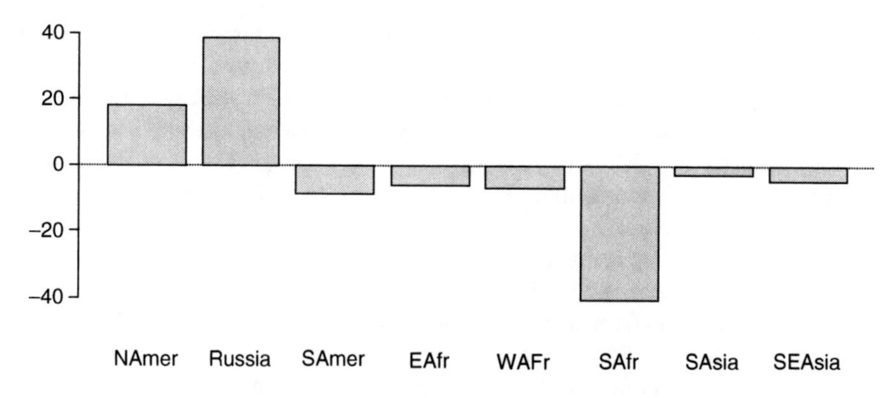

Fig. 8.3 Percent change in land suitability for rainfed cereal production, for selected regions by 2080 (Hadley model, A2 scenario) (from Fischer et al. 2002)

But in some systems irrigation may represent a truly autonomous adaptation, for instance if a treadle pump is installed to irrigate a small number of fields.

There is also considerable scope for implementing technologies that improve soil moisture without irrigation, such as conservation tillage and rainwater harvesting (Ngigi et al. 2005; Hobbs et al. 2008). The latter includes techniques such as farm ponds and zai pits, and may be increasingly relevant if rainfall becomes more episodic and intense, as suggested by many climate models (Chapter 3).

In areas currently too cold or dry to support rainfed agriculture, climate change might enable the expansion of cropped area into new regions. If such expansion is deemed socially and environmentally acceptable, then gains from production in these new areas could offset potential regional or global losses elsewhere (see Section 8.5).

Figure 8.3 shows one estimate of regional changes in the amount of land suitable for rainfed production, based on the agro-ecological zoning (AEZ) model and output from one climate model (Fischer et al. 2002). High latitude temperate regions generally gain and tropical areas generally lose suitable land in these projections, with changes exceeding 40% in either direction by the end of the century for some climate scenarios. Critical uncertainties in these projections are assumptions about soil constraints in these new regions, which are usually incorporated into assessments but on the basis of scant data. Improving the accuracy and use of soil information in these regions is a major need for determining future potential of expansion in places like Canada and Russia.

8.4.4 *Diversify Income*

On-farm adaptations are not the only possibility for bolstering food security in the face of a changing climate. Recall from Chapter 2 that while many rural poor lean heavily on agricultural activities for income generation, off-farm income can also play an important role in economic livelihoods. To the extent that non-agricultural income sources are less climate-sensitive than farm activities, further diversification of incomes out of agriculture might seem a promising adaptation strategy in the face of a changing climate. Indeed, some commentators have suggested that such a strategy is the only plausible way that Africa can adapt to climate change (Collier et al. 2008).

The ability of an income diversification strategy to buffer food security in the face of a short-run climate shock or longer-run climate shift depends on the off-farm income-generating activities available, and the extent to which households can take advantage of them. As Davis et al. (2007) show, almost all rural households earn at least some off-farm income, but the nature and motivation of this earning can differ significantly. For some households, off-farm work in manufacturing or in the service sector can offer much higher returns than farming, and households that can take advantage of these opportunities often benefit greatly.

But for many of the poorest households, participation in these potentially more lucrative non-farm activities is often limited by liquidity or human capital constraints (the cash to invest in a sewing machine, for instance, or the skills to run it). For these households, off-farm income generation often entails lower-return activities such as seasonal wage labor, which are used more as a coping strategy to deal with seasonal credit constraints in agriculture or with farm productivity shocks due to climate or other factors.

Using off-farm income as a climate coping strategy is likely more successful when climate shocks are idiosyncratic rather than covariate – i.e. when in a given year they affect some households in a region but not others. This is because in many developing countries, particularly the poorest ones, returns to off-farm activities can be highly correlated with agricultural productivity (Jayachandran 2006; World Bank 2008b). If most people in a village are farmers, and all experience a yield (and thus income) decline simultaneously, then demand for both agricultural wage labor and off-farm goods and services will likely also fall.

Overall, if there are specialization options available, and households can take advantage of them, then diversification looks like a very appealing adaptation to climate change. But where diversification is used as a necessary but low-return coping strategy and households face significant barriers to entry into higher-return activities, or where the non-farm rural economy is tightly linked to an agricultural sector deeply harmed by climate change, then income diversification looks less promising. Again, as with new technology adoption, diversification will likely be more challenging in poorer countries with less developed infrastructure, and for poorer households within those countries.

8.5 Broader Economic Adjustments to Climate Change

Even if individual farmers do not successfully perceive and adapt to climate change, market forces will tend to favor those farmers and regions that are more successful in the new climate. These market-mediated responses can range from individual farmers taking over their neighbor's land, to entire regions shifting into and out of production of different crops.

Most studies of market effects to date have focused on the latter mechanism, namely markets adjusting through international trade. All countries participate to some degree in international trade in agricultural commodities, and few households anywhere are fully isolated from markets. Under current climate variability, in which climate shocks typically correlate poorly across regions in a given year, global and regional agricultural markets can move food from areas of surplus to areas of deficit and dampen what might have otherwise been large price effects in regions experiencing shortfall.

Studies that attempt to directly capture these trade effects in understanding the potential impacts of climate typically embed regional production effects in a global trade model, which add up supply and demand across regions for a given period and calculate a market-clearing world price. Farmers and consumers then react to this price in the next period by adjusting what they produce and consume, new production effects are included, and a new world price calculated.

Such studies typically find that allowing countries to trade with one another tends to reduce the estimated negative impacts on global production, as production shifts into areas where the climate becomes more favorable (Rosenzweig et al. 1993; Darwin et al. 1995; Fischer et al. 2002). Figure 8.4 shows production impacts with and without economic adjustment estimated as reported by two major studies (also plotting estimates of gains from all farmer adaptations added together), which suggest that including these adjustments reduces estimated climate losses by anywhere between 25% and 75% of the unadjusted losses. These gains in turn dampen what would otherwise have been large increases in food prices, and reduce negative impacts on food security relative to a non-adjusting world.

But there are many important caveats to these conclusions that relate to the often poorly tested assumptions of the trade models. Most notably, growth in national GDP in these studies is often assumed to be independent of agricultural productivity

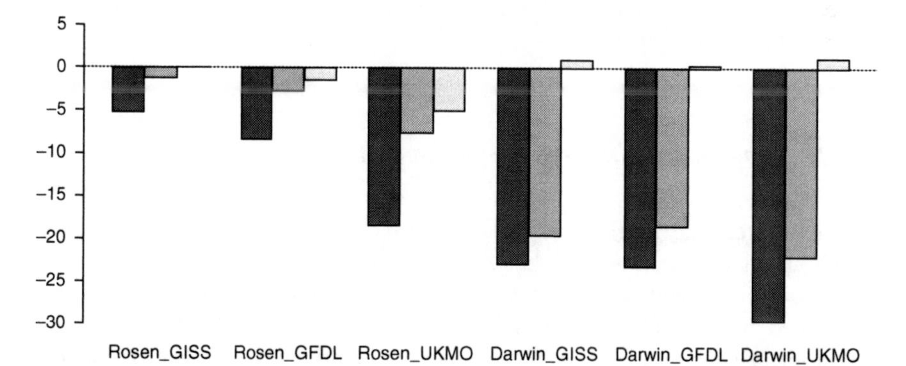

Fig. 8.4 Estimated effects of climate change on global cereal production to 2060 for two global studies, each running three climate models. *Dark grey* = no adjustment, no farmer adaptation; *medium grey* = economic adjustment, no farmer adaptation; *light grey* = with adjustment and farm-level adaptation (from Darwin et al. 1995 and Rosenzweig et al. 1993)

changes, and is projected into the future at rates often much higher than recent historical experience. As a result, declines in agricultural productivity do not translate into income declines, and so agriculturally dependent countries that are hit hard by climate change still have the income to purchase imports and cover production shortfalls, thus perhaps underestimating the income-related impacts on food security.

Whether or not agriculturally dependent countries (or households) will in fact be able to maintain food consumption in the face of declines in a primary income source is a crucial question, and underscores the importance of climate interactions with broader economic trends. On the whole, wealthier societies and households appear more adaptable to climate change: they are more willing to adopt higher-risk higher-return technologies because they can smooth consumption through savings or credit markets, they are less sensitive as consumers to food price rises, and they have the infrastructure and resources to import in the face of shortfalls.

Recall from Chapter 2 that climate is only one of many possible factors that shape a given country's longer-run economic trajectory. If households or societies are able to enrich themselves despite the potential adverse effects of climate change, then food security could overall become less sensitive to climate. But in countries where agriculture is a primary engine of growth, climate change could slow overall growth trajectories and limit the expansion of choice that typically accompanies economic development.

8.6 Planned Adaptations

Although autonomous adaptations of farmers and markets will certainly help, many studies indicate that they will be limited in their capacity to reduce the costs and impacts of climate change (Rosenzweig et al. 1993). Planned interventions by governments and other institutions may therefore be needed beyond what can be expected automatically. Here we provide a brief discussion of several potentially important planned adaptations.

8.6.1 Investments in Crop Development

As climate change pushes regional climates outside of historical experience, development of crop varieties better suited to these new climates will be an important component of adaptation. Chapter 9 reviews the breeding challenges associated with developing crops for new climates. Throughout much of the world, these challenges will mostly be met by the private sector. In high-income countries, the private sector accounts for 55% of total agricultural R&D expenditures, and many companies are actively publicizing their efforts to develop varieties well suited to changing climates (see, for instance, Monsanto's efforts with drought-tolerant maize).

But private sector investment will likely not be enough in many developing countries, where input markets are more poorly functioning and poor farmers represent limited economic demand for new varieties. Public-sector expenditures currently account for 94% of agricultural R&D in the developing world (Pardey et al. 2006), and historically these investments have yielded extremely high social returns (Alston 2000). Unfortunately, inflation-adjusted public sector spending on agricultural R&D in developing countries has been roughly stagnant since the 1980s, and key sources of external aid to developing country agriculture have fallen dramatically over the same period (Pardey and Beintema 2002). At the same time, however, large recent investment in agricultural development by foundations such as the Gates and Rockefeller are beginning to fill some of the public-sector void, particularly in Africa.

More broadly, given the decade or more it typically takes to develop and release new varieties, breeding programs face the difficult task of identifying regional and global priorities in the context of rapidly warming temperatures and continued uncertainty about the relative impacts of climate change on yields of different crops (Lobell et al. 2008). Supplying breeders with better information on the conditions and constraints that climate change will pose for future agricultural systems is therefore a major research priority.

8.6.2 Making Markets Work for the Poor

Developing improved agricultural technology will almost certainly be necessary for adapting agriculture to climate change, but it is unlikely to be sufficient. Current adoption of improved cereal varieties differs widely across Africa, with estimates ranging from 0% adoption of improved millet varieties across much of the continent, to 80% adoption of improved maize varieties in parts of East and Southern Africa (Maredia et al. 2000; World Bank 2008). To adapt to climate change, farmers need access to these improved technologies and the knowledge and incentives to use them. While information provision to farmers will likely continue to require direct public-sector action (see Section 8.6.3), farmer access to new technologies is likely better served by the private sector in the long run, given the high fiscal and administrative costs often associated with government input distribution programs (World Bank 2008). Governments are often better positioned to provide investments in the physical and financial infrastructure that underpin functioning agricultural markets. These could include investments in transportation infrastructure to better link farmers to input and output markets, investments in the functioning of these markets themselves, and investments in improving poor farmer access to financial infrastructure such as credit and insurance.

For instance, input markets in many poor regions – notably Africa – are often poorly functioning and hamper farmer response to changes in climate. Expanding private-sector provision of inputs like seeds and fertilizer faces numerous difficulties, including high transport costs and weak demand from credit constrained and risk averse farmers. Government investment in roads and ports could help reduce transport costs, and recent foundation investments in agrodealer networks in East Africa has shown promise in linking smallholders to input markets (World Bank 2008).

Similarly, improvements in financial infrastructure could boost both ex-post and ex-ante adaptation capabilities of farmers. Expanding the availability of credit and insurance in poor countries, for instance, could help farmers finance the purchase of inputs, smooth incomes in the face of production shortfalls, and thus encourage diversification out of low-risk, low-return crops and into higher-reward activities.

In particular, there is widespread interest in the development of crop insurance schemes that would reimburse farmers in the event of a climate-related production shortfall. If risk avoidance explains much of why poor farmers are reluctant to adopt higher-return technologies, then the availability of insurance could speed the adoption of new, better-adapted varieties, in addition to helping maintain incomes in bad years.

Providing climate insurance products to poor producers faces a number of hurdles, including the transaction costs of dealing with high numbers of dispersed smallholders, moral hazard problems (were observed production shortfalls a result of bad weather or farmer laziness?) and issues related to the covariate nature of climate risk. This latter concern, in which climate shocks cause simultaneous losses across farmers in a region and thus exceed the reserves of the insurer, is a primary explanation for why insurance is unavailable in many poor regions (Barnett et al. 2008). If climate change greatly increases the incidence of "bad" years, the stability of existing insurance schemes could be further compromised.

Various solutions have been proposed to overcome these problems, including the development of index-based insurance products where payouts are linked to a publicly observable index such as rainfall. In these products, payments would be triggered if rainfall (or some other variable) fell below a pre-determined threshold. Such "weather-indexed" crop insurance schemes would overcome moral hazard problems, and could be helped to remain solvent in the face of covariate shocks if further guaranteed by governments or larger financial institutions. Various products are being piloted throughout the developing world, with some apparent successes (World Bank 2005; Gine et al. 2008).

8.6.3 Building Local Knowledge

Public-sector involvement in information provision to farmers has long been a cornerstone of agricultural development strategies, with large proven benefits to agricultural output in both rich and poor countries (Birkhaeuser et al. 1991; Alston 2000). These strategies can involve educating farmers about the availability of new technology and how to use it, providing information on improved farm management techniques such as optimal input use, or providing forecast information about likely short- or longer-run shifts in climate. Including farmers in research design and implementation can also be an important means toward successful technology adoption. For example, adoption of new wheat varieties and no-till management in South Asia has been greatly accelerated through participatory research trials conducted in farmers' fields, where farmers' can see first-hand the benefits of new seeds or techniques (Ortiz-Ferrara et al. 2007).

8.6.4 Expansion of Irrigation Infrastructure

Irrigation was discussed above (Section 8.4.3) as a possible autonomous adaptation, but in many cases major public investments will be needed to provide farmers access to water. Some of these investments would undoubtedly happen even without climate change. For example, as part of its recent outlook assessment, the FAO projected changes in irrigated area for 93 developing countries notwithstanding climate change (Faurès et al. 2002). Overall an additional 40 Mha in irrigated area was anticipated by 2030, an increase of 20% over 1997–1999 levels. An increase in the cropping intensity (number of crops per year) on these lands is also anticipated, which results in a 33% increase in the effective area of crops harvested from irrigated land. A regional breakdown of these projections (Table 8.2) shows that most of the expansion in absolute terms is expected in Asia, with Africa anticipated to remain with only roughly 2% of cropland area under irrigation.

The additional irrigated areas will reduce impacts of climate change relative to no expansion, and in that sense will represent an adaptation. But as with most other planned adaptations, these investments also accrue benefits in the current climate, and some level of investment would therefore occur even without concern for climate change. Partitioning out the additional investments needed or benefits occurring because of climate change can therefore be difficult. This is similar to the questions of additionality that plague funding of mitigation projects, and will certainly be a challenge for evaluating pledges of adaptation funding in the future.

Nonetheless, it is clear that only irrigation beyond this baseline amount can truly be considered an explicit response to the added pressures of climate change. What will such investments cost? A recent review of project costs by the African

Table 8.2 One study's projection of increases in irrigated area for developing countries, without adaptation to climate change (Faurès et al. 2002)

	Irrigated area in 1997–1999		Irrigated area in 2030		Increase 1999–2030	
Region	Mha	As % of total crop area	Mha	As % of total crop area	Mha	%
All developing countries	202	21	242	22	40	20
Sub-Saharan Africa	5.3	2	6.8	2	1.5	28
Near-East/North Africa	26	30	33	35	7	27
Latin America and Carribbean	18	9	22	9	4	22
South Asia	81	39	95	44	14	17
East and Southeast Asia	71	31	85	36	14	20

Development Bank and the International Water Management Institute (Inocencio et al. 2007) puts the average cost of new irrigation projects at roughly $8,200/ha in developing countries, with higher costs in Sub-Saharan Africa ($14,500) relative to other regions (ranging from $3,400 in South Asia to $8,800 in the Middle East and North Africa). Much of this difference can be attributed to the smaller size of most irrigation projects in Africa, which increases per area costs.

Applying these costs to the expected rates of expansion in Table 8.2 yields a total cost of roughly $300 billion over the 30-year period. If doubling the anticipated rate is considered as a target for adaptation, then the cost would be roughly $10 billion per year. Doubling rates in Sub-Saharan Africa would cost roughly $650 million per year assuming past costs, although several strategies for cost reduction have been identified (Inocencio et al. 2007). These are of course extremely crude estimates, but they raise important questions about the opportunity costs of such investments, particularly given the dismal past performance of most large-scale irrigation projects in Africa (World Bank 2008). Potentially more cost-effective solutions include the rehabilitation of existing systems, investments in rainwater harvesting approaches (discussed in Section 8.4.3), and investments in smaller-scale irrigation systems for high-value crops.

8.6.5 When Adaptation in Agriculture Is Not Enough

Even if all of the above adaptation measures are taken (perhaps a big if), food systems may still not be fully shielded from the negative effects of a changing climate. As a result, a final set of planned adaptations might involve strengthening social safety nets to deal with climate-related shocks to food systems when they inevitably occur.

The expansion of insurance products to farmers (explored above) would be a primary means for smoothing producer income in the face of climate induced productivity shortfalls. But what about agricultural wage laborers whose incomes typically fall in bad climate years (Jayachandran 2006), and rural and urban net-consumers who are hurt by rising food prices? Typical social safety nets in this context include public works programs that employ individuals who would otherwise lose significant income in the face of a climate shock; conditional cash transfer schemes, in which payments are made to households in the face of a shock, conditional on some behavior (e.g. sending their children to school); or food aid, where donors contribute either food or cash, which is then distributed to households (in the case of direct food aid) or used by various organizations to purchase food locally which is then distributed.

Operation of these safety nets is typically improved when programs are in place before a shock arrives, and when governments hold reserve funds for their operation (given that government revenues, and thus funding, can also decline in a bad year) (World Bank 2008). In the specific case of food aid, most research suggests cash-based food aid is a more efficient means of aid delivery in the face of shortfalls, although there are caveats (Barrett and Maxwell 2005).

8.7 Measuring Progress in Adaptation

Given the importance of climate adaptation to the future of agriculture, it is imperative that we improve our understanding of how and how fast management and technologies adaptations will proceed. In particular, understanding the pace and impact of autonomous adaptation will be necessary for identifying the scope and type of needed planned adaptations. The recent and ongoing changes in climate may offer some insight into what farmers are actually doing in response. However, how will we recognize adaptation if and when it is happening? Among the many changes sure to occur in agricultural management and technology, will we be able to distinguish those that qualify as adaptation? Put more simply, what will an "adapted" food production system look like?

Broadly speaking, an adapted world in 2050 will have some key characteristics to look for: widespread planting of new crop varieties; area expansion of crops and shifts in planting dates, particularly in temperate regions; expansion of irrigation and water harvesting; and effective institutions for anticipating and responding to droughts and local food production shortfalls. Realizing this adapted world, however, will require difficult decisions on the part of public and private sector agencies around the world with regard to how, where and when to invest. Further scientific research will be critical in informing this process, both to further reduce uncertainties surrounding likely impacts in the absence of adaptation, and to identify regions where producers and consumers will be unable to respond on their own and where investment could be most needed.

8.8 Summary

The rapid pace of climate change and its anticipated large negative effects on many agricultural systems suggest a broad and pressing need for adaptation. For farming households, the nature of these responses will depend on their recognition that climate is changing and their ability to adjust their behavior in response, perhaps through altering farm management practices or diversifying into off-farm income-generating activities. Such responses must happen in the context of climate variability, which can obscure longer-run climate trends and make more risky the adoption of various adaptation measures. Further contributing to the difficulties is the limited choice set already faced by many food insecure households, which is often a result of high productivity risk, lack of access to insurance and credit, and/or limited connection to functioning input and output markets.

As a result, broader public and private investments will almost certainly be needed to help poor households adapt to climate change. These could include direct investments in the productivity of agriculture, such as in the development of improved crop varieties better suited to new climates; investments aimed at improving the physical and market infrastructure that typically underpin functioning economies; or investments that bolster the social safety nets that help poor households maintain their welfare in the face of a livelihood shock. While the optimal composition of investments will likely vary by country, scientific research can contribute important information concerning where climate change will hit hardest, how agricultural systems are likely to respond, and what particular investments in adaptation could yield high returns.

References

Alston J M (2000) A meta-analysis of rates of return to agricultural R&D: ex pede herculem? Int Food Policy Res Inst IFPRI

Barnett BJ, Barrett CB et al (2008) Poverty traps and index-based risk transfer products. World Dev **36**(10):1766–1785

Barrett CB, Maxwell DG (2005) Food aid after fifty years: recasting its role. Routledge, London

Battisti D, Naylor RL (2009) Historical Warnings of future food insecurity with unprecedented seasonal heat. Science **323**(5911):240

Beinhocker ED (2006) The origin of wealth: evolution, complexity, and the radical remaking of economics. Harvard Business School Press

Birkhaeuser D, Evenson RE et al (1991) The economic impact of agricultural extension: a review. Econ Dev Cult Change **39**(3):607–650

Cline, William (2007) Global Warming and Agriculture: Impact Estimates by Country. Peterson Institute, Washington DC, 201 pages

Collier P, Conway G et al (2008) "Climate change and Africa. Oxford Rev Econ Pol 24(2):337

Cooper PJM, Dimes J et al (2008) Coping better with current climatic variability in the rain-fed farming systems of sub-Saharan Africa: an essential first step in adapting to future climate change?" Agric Ecosyst Environ **126**(1–2):24–35

Darwin R, Tsigas M et al (1995) World Agriculture and climate change: economic adaptations. Economic Research Service, US Department of Agriculture

Davis B, Winters P et al (2007) Rural income generating activities: a cross country comparison. ESA Working Paper 68. FAO, Rome

Dercon S-R (2002) Income risk, coping strategies, and safety nets. World Bank Research Observer 17(2):141–166

Falcon WP, Naylor RL et al (2004) Using climate models to improve Indonesian food security. Bull Indones Econ Stud 40(3):355–377

Faurès J, Hoogeveen J et al (2002) The FAO irrigated area forecast for 2030. FAO:14, Rome, Italy

Fischer G, Shah M, van Velthuizen, H (2002) Climate change and agricultural vulnerability, International Institute for Applied Systems Analysis, Vienna, 160 pages

Fisher A, Hanemann M et al (2007) Potential impacts of climate change on crop yields and land values in US agriculture: negative, significant, and robust. University of California, Berkeley

Gine X, Townsend R et al (2008) "Patterns of rainfall insurance participation in rural India." World Bank Econ Rev 22(3):539

Hanemann WM (2000) "Adaptation and its measurement." Climatic Change 45(3):571–581

Hobbs PR, Sayre K et al (2008) The role of conservation agriculture in sustainable agriculture. Philos Trans R Soc B: Biol Sci 363(1491):543–555

Hoddinott J, Kinsey B (2001) "Child growth in the time of drought." Oxford Bull Econ Stat 63(4):409–438

Inocencio A, Kikuchi M et al (2007) Costs and performance of irrigation projects: a comparison of Sub-Saharan Africa and other developing regions. International Water Management Institute, Colombo, Sri Lanka

Jayachandran S (2006) Selling labor low: wage responses to productivity shocks in developing countries. J Polit Econ 114(3):538–575
Kandlikar M, Risbey J (2000) Agricultural impacts of climate change: if adaptation is the answer, what is the question? Climatic Change 45(3):529–539
Lobell DB, Ortiz-Monasterio JI et al (2005) Analysis of wheat yield and climatic trends in Mexico. Field Crop Res 94(2–3):250–256
Lobell DB, Burke MB et al (2008) Prioritizing climate change adaptation needs for food security in 2030. Science 319(5863):607–610
Maccini S, Yang D (2008) Under the weather: health, schooling, and economic consequences of early-life rainfall. NBER Working Paper
Maddison D (2007) The perception of and adaptation to climate change in Africa. World Bank, Washington, DC
Maredia MK, Byerlee D et al (2000) Impacts of food crop improvement research: evidence from sub-Saharan Africa. Food Policy 25(5):531–559
Mendelsohn R, Nordhaus WD et al (1994) The impact of global warming on agriculture: a Ricardian analysis. Am Econ Rev:753–771
Meze-Hausken E (2004) Contrasting climate variability and meteorological drought with perceived drought and climate change in northern Ethiopia. Climate Res 27(1):19–31
Ngigi SN, Savenije HHG et al (2005) Agro-hydrological evaluation of on-farm rainwater storage systems for supplemental irrigation in Laikipia district, Kenya. Agric Water Manage 73(1):21–41
Ortiz-Ferrara G, Joshi A et al (2007) Partnering with farmers to accelerate adoption of new technologies in South Asia to improve wheat productivity. Euphytica 157(3):399–407
Pandey S, Bhandari HS et al (2007) Economic costs of drought and rice farmers' coping mechanisms: a cross-country comparative analysis. International Rice Research Institute
Pardey PG, Beintema NM (2002) Slow magic: agricultural R&D a century after Mendel. International Food Policy Research Institute, Washington, DC
Pardey PG, Beintema N et al (2006) Agricultural research: a growing global divide?. IFPRI, Washington, DC
Reilly J, Schimmelpfennig D (2000) Irreversibility, uncertainty, and learning: portraits of adaptation to long-term climate change. Climatic Change 45(1):253–278
Rosenzweig MR, Binswanger HP (1993) Wealth, weather risk and the composition and profitability of agricultural investments. Econ J 103:56–78
Rosenzweig C, Parry M et al (1993) Climate change and world food supply. University of Oxford, Environmental Change Unit, Oxford
Smit B, Blain R et al (1997) Corn hybrid selection and climatic variability: gambling with nature? Can Geogr 41(4):429–438
Stefan Dercon (2002) Income risk, coping strategies, and safety nets. World Bank Research Observer 17(2):141–166
Tadross MA, Hewitson BC et al (2005) The interannual variability of the onset of the maize growing season over South Africa and Zimbabwe. J Clim 18(16):3356–3372
Thomas DSG, Twyman C et al (2007) Adaptation to climate change and variability: farmer responses to intra-seasonal precipitation trends in South Africa. Climatic Change 83(3):301–322
Tubiello FN, Rosenzweig C et al (2002) Effects of climate change on US crop production: simulation results using two different GCM scenarios. Part I: wheat, potato, maize, and citrus. Climate Res 20(3):259–270
Washington R, Harrison M et al (2006) African climate change: taking the shorter route. Bull Am Meteorol Soc 87(10):1355–1366
World Bank (2005) Managing agricultural production risk: Innovations in developing countries. World Bank, Washington, DC
World Bank (2008) Investment in agricultural water for poverty reduction and economic growth in sub-Saharan Africa. World Bank, Washington, DC
World Bank (2008) World development report: agriculture for development. World Bank: 386, Washington, DC

Chapter 9
Breeding Strategies to Adapt Crops to a Changing Climate

R.M. Trethowan, M.A. Turner, and T.M. Chattha

Abstract Climate change is expected to reduce global crop productivity, although the impact will vary region to region. At many locations, particularly those at lower latitudes, the environment will become drier and hotter, which will reduce crop yields and potentially change the incidence of insect pests and diseases. These climatic changes are also expected to alter the nutritional properties and processing quality of crop products. This chapter describes breeding approaches that may be employed to mitigate the effects of increased heat and drought in the crop production environment.

9.1 Introduction

Predictions of climate change have different consequences for crop production globally. In some instances environments will become drier and hotter, in others precipitation will increase and rising temperature will expand the scope of crop production, particularly at higher latitudes (Christensen et al. 2007). However, the negative impact of climate change will likely be far greater closer to the equator, in some of the world's poorest and most densely populated countries. Forecasts indicate that elevated CO_2 levels will have a fertilizing effect in some regions, although this will be negated by greater drought and heat stress in lower latitude areas. In most developing countries, wheat, rice and maize are the primary source of calories for the vast majority of people and any fall in production could have dire humanitarian consequences. At the same time, most commentators estimate that global production of food grains must double by 2050 to keep pace with increasing population and demand for food (APA 2004).

R.M. Trethowan (✉), M.A. Turner, and T.M. Chattha
The University of Sydney, Plant Breeding Institute, PMB 11 Camden, NSW, 2570, Australia

D. Lobell and M. Burke (eds.), *Climate Change and Food Security*,
Advances in Global Change Research 37, DOI 10.1007/978-90-481-2953-9_9,

Under these circumstances, agricultural scientists have two main options to increase the productivity of agriculture: development of better management practices, and development of better agricultural technology. While improved understanding of best management practices such as conservation agriculture have greatly improved food production systems in much of the world, improved agricultural technology – specifically the development of more water efficient cultivars with improved heat tolerance – offer some of the greatest hope for improving crop productivity in an increasingly hostile environment.

This chapter will focus on genetic options that can be used to improve the water-use-efficiency and heat tolerance of crop cultivars. Successful breeding depends on four sequential steps, which we review below: (1) identifying the target traits, which are a function of the target growing environment, (2) identifying sources of genetic variability for these traits; (3) crossing these sources of variability with existing varieties that possess other traits of economic importance such as disease resistance and high yield or quality; and (4) testing these new varieties across a range of on-farm environments. Wheat will be used as the model species, but the same principles apply for rice, barley and many other small grained cereals and to a large extent to open pollinated species such as maize.

9.2 Breeding Wheat for Adaptation to Moisture Stress and Increased Temperature

The first step in breeding crops with improved response to water and temperature stress is identification of genetic variability governing the plant response. This response may be environment specific and its genetic control is likely to be complex. Soil type and associated water holding capacity and infiltration rates, crop management practices, timing of water stress during the plant growth cycle, temperature and biotic constraints will all influence plant response to drought. Determination of the dominant stress patterns in the target environment is of critical importance if the appropriate genetic variability is to be identified and used in crossing. Chapman et al. (2003) used the concept of 'target population of environments' (TPE) to identify dominant environment types in space and time. Historic weather data, genotype performance in multi-environment trials (METs) and crop distribution information can be used to determine the frequency of occurrence of defined stresses (Edmeades et al. 2006). Once these are known, weighting can be given to those locations representing the TPE in any MET analysis, thereby improving the breeder's selection of appropriate germplasm. Once the TPE is identified and the underlying environmental constraints understood, it will be possible to select parents representing the genetic variability needed to improve adaptation in the TPE. If climate modeling indicates a change in the dominant stress pattern over the next 10–20 years, the breeder can give weighting to the occurrence of this future TPE in the MET analysis, thereby skewing gene frequency in favor of adaptation to the predicted conditions in the target region. It may be possible to develop managed selection environments that mimic the TPE and further improve the selection response.

Climate predictions for the state of New South Wales in Australia typify the challenges crop breeders face in targeting their long-term breeding objectives. Projections for the year 2030 are that the frequency of drought will increase by 70% in the worst case scenario (decreased rainfall) and decrease by 35% in the best case scenario (increased rainfall) (Hennessy et al. 2004). Given these predictions, it is sensible for a plant breeder to assume that improved drought and heat tolerance will be beneficial in the future production environment. Chapter 3 presents more specific information on expected rates of heat and drought changes in key agricultural regions.

9.2.1 Genetic Variability

9.2.1.1 The Breeding Program Gene Pool

The genetic constitution of wheat is both tetraploid (i.e., containing four sets of chromosomes, as in the case of *Triticum turgidum* or durum wheat) and hexaploid (i.e., containing six sets of chromosomes, as in the case of *Triticum aestivum* L. or bread wheat), and this presents both opportunities and difficulties for its improvement. Durum wheat is a fusion of two diploid species and its genetic constitution is denoted as AABB, whereas bread wheat originated from a cross between tetraploid AABB species and a third diploid species to produce a hexaploid AABBDD constitution.

Diploid species (with two sets of chromosomes) such as rice and barley carry less diversity both within the cultivated gene pool and among the ancestral species, but they are more easily manipulated genetically. Once the TPE has been defined the breeder must then identify genetic variability conferring improved adaptation to this dominant stress pattern. In some instances there may be more than a single dominant stress in the target region. The first exercise is to identify those materials within the breeding program gene pool with superior performance within the TPE; these are the backbone of the crossing strategy. There are a number of excellent options available for the analysis of MET data, including but not restricted to cumulative cluster analysis that estimates the association among sites and genotypes using unbalanced MET data (DeLacy et al. 1996) and the Shifted Multiplicative Model and Sites Regression model for the analysis of balanced data (Crossa et al. 1993).

9.2.1.2 Landrace Cultivars

The Green Revolution resulted in massive increases in wheat and rice production globally which to some extent narrowed genetic variability as farmers adopted high yielding, short-statured cultivars in most production environments (Warburton et al. 2006). However, farmers in marginal areas did not adopt these modern cultivars at the same rate (Byerlee and Moya 1993). This reduced adoption reflects the risk-averse nature of the farmers in these marginal cropping lands. The higher moisture and temperature stresses that are characteristic of these environments have also lessened the impact of the high yielding, resource responsive germplasm. Landrace collections existing either in situ or in gene banks around the world present a potential source of genetic variability for the improvement of stress tolerance. The first step in their utilization is to screen available collections for response to both drought and heat stress typical of the TPE. However, GIS (geographic information system) tools have improved the efficiency with which this can be done. Instead of screening thousands of lines, information on the geographic location and associated environmental conditions under which the germplasm was collected can be used to identify materials likely to adapt to the TPE (Greene et al. 1999).

Landraces have been found to be more water-use–efficient, extracting water from deeper in the soil profile than modern cultivars and possessing higher soluble stem carbohydrates (Reynolds et al. 2007a, Reynolds and Trethowan 2007). They are also more heat tolerant, characterized by higher leaf chlorophyll and higher stomatal conductance (Hede et al. 1999; Skovmand et al. 2001). One might argue that the stress tolerance that is present in the landraces is the same as that found in modern cultivars as the modern materials were derived from landraces. However, genotyping studies show that stress tolerant landraces are generally genetically distant from the more tolerant modern wheats (Moghaddam et al. 2005; Reynolds et al. 2007b). Similarly, in a study of landrace diversity in the backgrounds of 143 commercial rice cultivars in Brazil, it was found that only 14 ancient cultivars contributed 70 percent of the important genes (Guimaraes 2002). Clearly, there is significant scope to broaden the genetic base of important crops using landraces.

9.2.1.3 Synthetic Hexaploid Wheat

As bread wheat is hexaploid, the opportunity exists to exploit variability among its progenitor species. Bread wheat likely arose from a cross between *Triticum dicoccom* and *Aegilops tauschii* following spontaneous chromosome doubling some 8,000–9,000 years ago. It is likely that very few accessions of both species were involved in this initial hybridization and subsequent evolution of wheat (Feldman 2001). Primary synthetic wheat can be generated in the laboratory from crosses between tetraploid wheat, either modern durum wheat (*T. durum L.*) or *T. dicoccum*, and *Aegilops tauschii*. These new hexaploids are agronomically poor, difficult to thresh and have poor end-use quality but carry unique genetic diversity. The resultant primaries have been screened for performance in the field under moisture deficit and high temperature stress and useful genetic variability found (Villareal and Mujeeb-Kazi 1999; Yang et al. 2002).

9.2.1.4 Alien Introgression

While the materials discussed so far have at least one genome in common with hexaploid wheat, significant genetic variation exists in the more distantly related tertiary gene pool (Trethowan and Mujeeb-Kazi 2008). The genomes of these materials do not recombine easily with wheat and are therefore difficult to exploit. However, alien chromosome segments have been introduced into wheat using the *ph* mutant which promotes their pairing (Sears 1976; Gupta et al. 2005). While most of the alien gene introductions todate target disease resistance, there has been useful variability reported for drought and heat tolerance. The replacement of the long arm of chromosome 1B with the short arm of rye chromosome 1R in wheat is probably the best example of alien introgression for both disease resistance and stress tolerance (Rajaram et al. 1983; Villareal et al. 1995). This translocation was found in the winter wheat cultivar Kavkaz and has been shown to increase root vigour and water up-take (Ehdaie et al. 2003). These distant relatives of wheat are a rich source of genetic variability.

9.2.1.5 Characters Important in Conservation Agriculture

As adoption of improved management practices around the world increases, crop cultivars better adapted to water and resource conserving farming practices will be important in improving the overall productivity of the farming system. Of importance to the breeder is the existence of a genotype x tillage practice interaction, as this will indicate whether or not breeding for specific adaptation to conservation agriculture is possible. There is evidence of genotype x tillage practice interactions in wheat for yield and product quality (Gutierrez 2006). However, evidence is conflicting across different crops with non-significant interactions reported for barley (Ullrich and Muir 1986), sorghum (Francis et al. 1986), rice (Melo et al. 2005) and soybean (Elmore 1990) and both significant and non-significant interactions reported for maize (Brakke et al. 1983; Newhouse 1985). The lack of significant interactions likely reflects the small number of genotypes examined in these studies and the fact that all the materials tested were developed under conventional or complete tillage.

It is useful to the breeder if genotype response to conservation agriculture can be broken down into individual traits for selection. Traits considered important in conferring adaptation to conservation agriculture include the length of the emerging shoot or coleoptile (Rebetzke et al. 2007; Trethowan et al. 2001a), coleoptile thickness (Rebetzke et al. 2004), emergence from depth (Trethowan et al. 2005), seedling vigor (Liang and Richards 1999), rate of stubble decomposition (Joshi et al. 2007), root depth (Reynolds and Trethowan 2007), allelopathy (Bertholdsson 2005), N-use-efficiency (Ginkel et al. 2001), disease resistance (Trethowan et al. 2005) and seedling temperature tolerance (Boubaker and Yamada 1991).

9.2.1.6 Root Disease Resistance

No discussion of available genetic variability to improve stress tolerance in wheat is complete without considering resistance to root diseases. In farming systems prone to root rots and nematodes, disease resistance can improve water-use–efficiency by maintaining a healthy root system (Govaerts et al. 2007). The inheritance of these traits is relatively simple, compared to drought and heat response per se, and resistance is therefore more easily manipulated.

Screening plants for resistance to these diseases in the field or green house is difficult and there are many misclassifications of resistance. These escapes or misclassifications result in a relatively low heritability or low repeatability of the screening procedures. However, molecular markers linked to the genes that confer resistance are available for a number of important traits and can be used to improve the efficiency of gene introduction (Okogbenin et al. 2007).

9.2.1.7 Nutritional Quality

Wheat is one of the world's most important sources of food. It is made into products as diverse as leavened bread, flat bread, steamed bread, noodles, biscuits and cakes and wheat starch is used as an additive in many processed foods. While both processing and nutritional quality is under genetic control, the expression of quality is greatly influenced by the environment in which the crop is grown. The environment includes soil type and fertility, crop management practices and the prevailing weather conditions during crop development.

Micronutrient deficiency in humans, which is caused by inadequate intake of elements such as zinc and iron, impairs normal development and increases the incidence of disease, particularly in children of developing countries (Ezzati et al. 2002; Kennedy et al. 2002; Welch and Graham 2004). Micronutrients are concentrated mainly in the seed coat and embryo of the wheat grain and only small amounts are present in the starchy endosperm (Ozturk et al. 2006), so yield increases alone will not substantially increase micronutrient intake. Refined flours, generated by removing the bran and germ fractions, contain substantially lower concentrations of micronutrients than wholemeal flours or grain. There is a trend towards increased consumption of manufactured products developed from refined flour in some developing countries (Pingali 2007). However, these changes are often associated with increasing affluence and the impact of this trend on human nutrition will to some extent be mitigated by improved access to other more nutritional foods.

The elevated temperatures and CO_2 and drier conditions predicted in some regions are unlikely to impact upon the nutritional status of the major crops (see also Chapter 7). In some instances the lower crop yields from these more hostile growing conditions may increase micronutrient concentrations as the ratio of endosperm to seed coat will reduce. However, the negative impact of significantly lower productivity will dwarf any perceived benefit. Variation in micronutrient concentration is present in various crop species (Reddy et al. 2005; Menkir 2008;

Murphy et al. 2008). The Fe and Zn concentration in wheat seed appears to be quantitatively controlled (Trethowan et al. 2005; Trethowan 2007) and molecular markers linked to a gene of major effect for Zn and Fe concentration have been reported (Uauy et al. 2006). Nevertheless, as micronutrient concentrations are higher in the seed coat compared to endosperm, improving the micronutrient concentration of the endosperm will be a key objective. Unfortunately, no substantial variation for endosperm Fe and Zn concentration is reported. In these instances it is likely that transgenic approaches will provide the only viable avenue for improving both yield and nutritional status.

Fungal mycotoxins can reduce the nutritional status of foods and in significant concentrations food can become dangerous to ingest. The production of toxins on cereal grain by fungi such as *Aspergillus* and *Fusarium* may increase in some regions as these organisms thrive at elevated temperatures and in conditions of plant stress (FAO 2001). There is genetic variation for resistance to these diseases in the wheat gene pool and *Fusarium* resistant wheat cultivars have been developed and deployed (Mergoum et al. 2006). However, the expression of resistance is generally incomplete and the quantitative nature of inheritance and low heritability makes breeding difficult (Jiang et al. 2007).

9.2.1.8 Product and Processing Quality

Most grain crops are consumed following processing of some sort. Small and shriveled wheat grains called screenings have reduced endosperm development, contain a higher proportion of bran and are more expensive to mill compared to normal grain. In developed countries grain with high levels of screenings is normally not used for food production but fed to animals instead. Screenings tend to increase when crops are subjected to water and/or temperature stress. An association between seed size and the incidence of screenings has been reported and there is scope to increase the seed size of wheat (Sharma and Anderson 2004).

Water and temperature stress can change the chemical composition of grain and its subsequent processing and product quality. In the context of food security, these changes are relatively minor as they largely affect the aesthetic appeal and cost of processing. However, in more advanced economies these affects take on greater significance and breeding for improved end-use quality under stress is important.

Water and temperature stress will alter the protein content and composition of wheat grain and subsequent end-use quality. High temperature is known to negatively impact gluten quality (Blumenthal et al. 1994) with subsequent effects on dough water absorption and the product quality of breads, biscuits, noodles and pasta. Proteins that have upregulated expression upon exposure to heat shock, and are thought to be associated with stability of quality when elevated temperatures occur during grain development, have been identified and partially characterized (Skylas et al. 2002). It appears that there is scope to breed for enhanced heat tolerance by selection for alleles of these proteins that are upregulated after heat stress.

Increased drought stress also enhances the yellow color of yellow alkaline noodles (commonly consumed in eastern Asia) but decreases their initial brightness, and can increase the grain hardness of soft-grained biscuit wheat (Guttieri et al. 2001; Weightman et al. 2008). There is considerable variation in the wheat gene pool for grain protein content and quality, grain hardness, strength and extensibility of dough and starch quality. However, the optimal balance of these properties and the genes that control their expression, in an increasingly variable and hostile growing environment must be determined for the breeder's TPE if realistic selection targets are to be set.

9.2.2 *Breeding Strategies to Improve Productivity and End-Use Quality Under Moisture Deficit and Higher Temperature*

A basic breeding scheme for a self-pollinated crop such as wheat is outlined in Fig. 9.1. This scheme represents either a modified bulk or selected bulk selection strategy. In a modified bulk strategy individual plants identified in the early generations are grown as individual plots in the following generation. These progeny are usually derived from three-way or top crosses (involving three parents) or simple crosses between two parents, and individual plants are usually selected from the

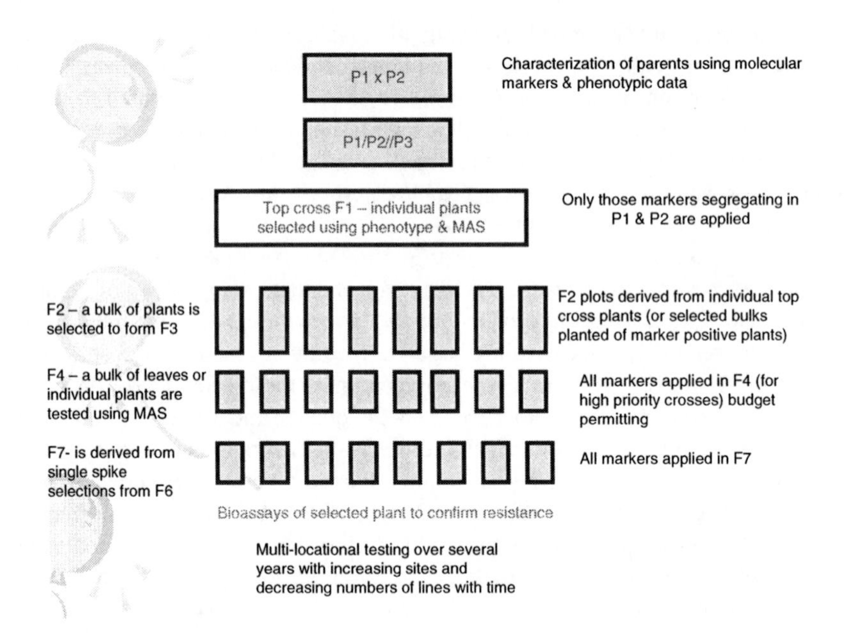

Fig. 9.1 An example of a conventional breeding scheme using either a modified bulk or selected bulk strategy. The time from cross to homozygous line identification is 4–7 years and a further 4–5 years of yield and quality evaluation and seed multiplication are required before the selected genotype is released to farmers

first segregating generation (either top cross F_1 or F_2). A bulk of selected spikes or plants is then used to advance the population to the next generation. This continues until de-bulking in the F_5 or F_6 generations to produce homozygous inbred lines.

A selected bulk differs in that selected F_2 plants are bulked to form a single F_3 bulk per cross. The populations are advanced in the same way until de-bulking in the later generations. The scheme in Fig. 9.1 assumes the use of molecular marker assisted selection (MAS) and drought and/or heat screening during the segregating phase (F_2–F_6). There are many variations on these schemes and the process described is one of among many possible strategies. The crossing, selection and evaluation strategies outlined in the following sections will be discussed in the context of the strategy in Fig. 9.1.

9.2.2.1 Crossing

Once genetic variability for adaptation to the prevailing stresses in the TPE has been assembled, the challenge for the plant breeder is to combine this variability in a crossing program that also encompasses the key biotic and market constraints. Genotyping technology has improved significantly in recent years and the breeder should have DNA fingerprints of all the key progenitors in the breeding program. Historically, breeders used coefficients of parentage that assumed no selection when determining relatedness among materials. The breeder will use this information to better design crosses. The degree of relatedness among parents selected for crossing will reflect the breeding objective, the complexity of the target trait and the available resources.

DNA profiles generated using microsatellites or DArT (Diversity array technology) (Mace et al. 2008) provide good genome coverage and offer more realistic estimates of diversity. Assuming that much of the variation for drought tolerance is additive (Trethowan and Mujeeb-Kazi 2008), the breeder can identify the least related lines from among the best performing materials under stress to combine in crossing. Association genetics studies may also be useful in identifying genomic regions linked to improved yield performance. Crossa et al. (2007) genotyped MET entries at the International Maize and Wheat Improvement Center (CIMMYT) spanning a 25-year period and related these profiles to wheat cultivar performance. Their analysis identified genomic regions unrelated to genes controlling phenology, morphology and disease resistance that were associated with superior yield in many environments globally. Combining these regions in crosses may provide the additive variance for stress response needed to improve the broad adaptability of crop germplasm. Characterization of parental materials is not just confined to MET data and molecular analysis. Determination of the physiological responses of progenitors within the breeding program to abiotic stress will allow the breeder to combine physiological mechanisms in crosses (Reynolds and Trethowan 2007). Progeny developed in this way at CIMMYT have shown superior performance in global MET experiments (Yann Manes, 2008 personal communication).

When introducing variability from primary synthetic wheat, a landrace or a translocation stock cultivar, it is in most instances sensible to make at least one backcross or top cross to an elite parent before proceeding to F_2. This is because most breeding programs cannot manage the extremely large populations required to exploit a simple cross between an adapted cultivar and a primary synthetic. In contrast, the backcross F_2 will produce a higher frequency of agronomically acceptable progeny. This principle was used at CIMMYT to produce synthetic derivatives with drought and heat tolerance, broad adaptation and high yield (Trethowan and Mujeeb-Kazi 2008).

In some instances the breeder may attempt to combine variability in double crosses or four-way crosses (crosses between F_1 progeny), although these usually result in F_2 progeny that are generally too variable to manage. However, should a reasonable degree of relatedness exist among two or more of the progenitors then such crosses may make sense. In reality the plant breeder gradually improves the frequency of favorable alleles in the breeding program over the span of a career and often several cycles of crossing and selection are required to pyramid genes for traits of economic importance.

9.2.2.2 Selection

Once the desired crosses have been made the selection of the segregating materials becomes vital. If molecular markers for known genes are available, they can be tracked in the segregating phase. Allele enrichment in the top-cross F_1, backcross F_1 and F_2 using markers for known genes will greatly increase the frequency of lines carrying the target genes in the subsequent fixed line progeny and can be useful for accumulating genes governing root health (William et al. 2007; Bonnett et al. 2005). If quantitative trait loci (QTL) of significant effect relevant to the TPE are available they can be introduced into elite germplasm using a MAS scheme similar to Fig. 9.1. However, significant QTL x environment interaction and genotype specificity tend to limit this approach. The breeder is often faced with multiple QTLs of relatively minor effect that are genotype and environment dependent. In this instance one possibility is to use a recurrent selection scheme, combined with molecular markers and empirical selection under stress, to provide a mechanism whereby these minor QTLs can be combined. In such a scheme parents would be genotyped using markers and QTLs combined in crosses and tracked using markers. The progeny would be genotyped and screened in multi-environment trials under stress at F_4 and the progeny selected on the basis of yield and genotype. These progeny would then be randomly intermated to continue the process of allele accumulation. This approach favorably skews gene frequency towards better adaptation under stress.

However, in the absence of QTLs for abiotic stress tolerance, favorably skewing gene frequency to greater levels of water or temperature stress tolerance will require one of two approaches. The segregating materials can either be selected in the TPE under all the prevailing stresses within any given year and site, or in managed selection environments that mimic the TPE. Effective selection in the TPE is dependent upon the occurrence of the TPE in the year of selection. The heritability of selection is extremely low for most water limited environments and year effects are almost always the largest component of variance (Ribaut et al. 1996; Ahmad and Bajelan 2008).

On the other hand, managed environments can increase the heritability of selection but their effectiveness is dependent upon correlation with the desired TPE. An analysis of global MET data of wheat lines developed for semi arid environments at the International Maize and Wheat Improvement Centre (CIMMYT) showed that the germplasm did not adapt to certain dry environments (Trethowan et al. 2001b). All the materials in this study were developed using simulated post-anthesis drought stress in the field in northwestern Mexico. However, the patterns of adaptation improved when a broader range of managed stresses were employed that better correlated with the target environment (Trethowan et al. 2005). The relevance of the managed selection environments at CIMMYT was confirmed in a retrospective study of genotypes previously tested in global METs in managed stress treatments. It was clear that specific stress patterns correlated with specific environments (Trethowan et al. 2005). However, given the vagaries of the target environment there will never be sufficient data to draw water-tight conclusions. This calibration of managed stress environments must be continual and an integral part of the breeding strategy.

While physiological characters, such as total soluble stem carbohydrates and transpiration efficiency are useful in differentiating parental materials and improving the efficiency of crossing for stress tolerance, their determination is generally not manageable in segregating generations where large numbers of lines have to be assessed. However, easy to measure physiological traits such as canopy temperature depression (CTD) do correlate with plant performance under both heat and drought stress (Reynolds et al. 2007a). If a modified bulk or selected bulk scheme is used, it is possible to measure CTD quickly and efficiently on large numbers of F_3 or F_4 plots (Reynolds and Trethowan 2007) (Fig. 9.2). The breeder's eye is the best physiological tool available, however CTD measured on breeder selected plots can show a significant range in temperature responses. When these materials were carried forward and fixed lines derived from them it is interesting to note that none of the

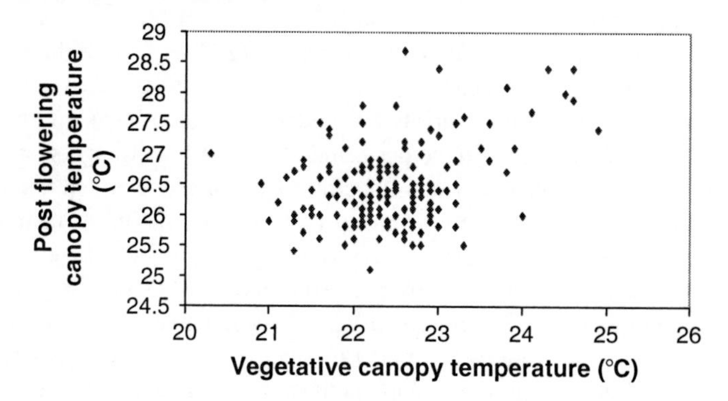

Fig. 9.2 The canopy temperature of F4 bulks under drought stress before and after flowering

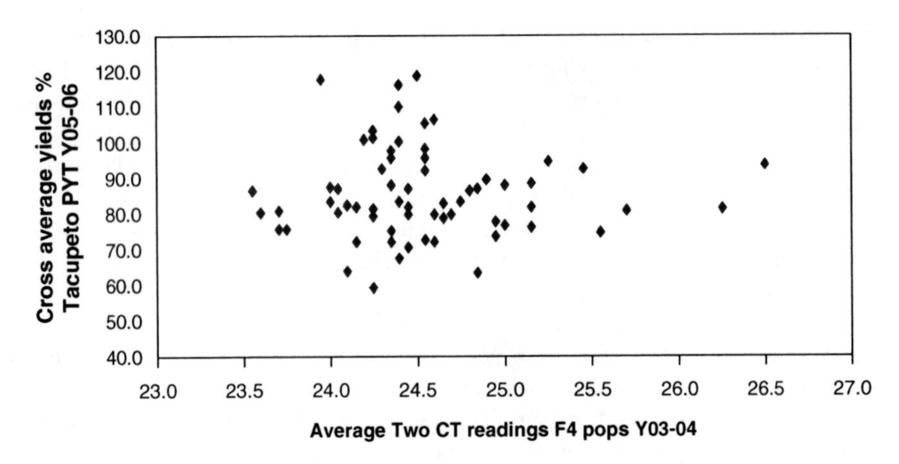

Fig. 9.3 The relationship between canopy temperature (CT) of F4 bulks in 2004 and the yield of fixed lines derived from them in 2006 (Yann Manes, unpublished data)

high-yielding lines had warmer canopies in the F4 generation (Fig. 9.3). This illustrates that CTD can be a reliable tool for identifying stress tolerant lines.

A similar approach can be used to select for end-use quality. A combination of molecular markers and high throughput small scale tests that correlate with end product quality can be used in the early generations to favorably skew gene frequency. This is possible where pedigree or modified bulk breeding strategies are used and seed from small plots are available. For example, these tests may be conducted on F_2 derived F_4 plots tested at more than one location.

As mentioned earlier, the deployment of more water-use–efficient and heat tolerant materials in conservation farming systems will improve overall productivity while conserving moisture and resources and reducing costs. Capturing the genetic response to these systems will require selection under zero-tillage with crop residue cover equivalent to that in the TPE. Alternatively, as crop emergence and establishment are important components of adaptation to conservation agriculture, planting segregating bulks deeper than normal combined with selection for short-statured plants from among those that emerge is an effective way of favorably skewing gene frequency (Trethowan et al. 2005). Trethowan et al. (2009) reported the results of a selection study in which segregating materials from the same cross were selected either always under zero-tillage or always under complete tillage. In general, the materials selected under zero-tillage performed better in both tillage systems.

The selection of some economically important traits such as elevated micronutrient concentration is hampered by the high cost of analysis. In these instances, the identification of linked molecular markers would greatly reduce the cost of selection. Molecular markers would ideally be used in the early segregating generations to skew gene frequency with subsequent ICP-MS (inductively coupled mass spectrometry) analysis of the relatively smaller number of fixed line progeny remaining at the end of the selection process. Markers have the advantage of being phenology independent, as differences in maturity within a population can confuse traditional selection approaches for traits such as Fe and Zn concentration. If the breeder has access to differential stresses (usually generated using limited irrigation or different planting dates) it would be informative to test the stability of expression of micronutrient concentration across two environmental extremes, typical of the current and predicted TPE.

In crops such as wheat and barley, double haploids provide an option for the rapid production of genetically stable homozygous lines. It is generally advisable to make the double haploids on F_2 or F_3 progeny once screening for simply inherited but economically important traits has been completed, thus greatly increasing the frequency of useable materials among the resultant double haploids. However, double haploids are unlikely to be particularly useful for the improvement of complex traits such as tolerance to water and temperature stress. Without selection under these stresses to improve allele frequency for plant response the probability of finding a double haploid with all the desired alleles is very small.

9.2.2.3 Evaluation

Once fixed lines have been identified, usually derived from single plants in the F5 generation or greater, the efficacy of the stress response must be confirmed. In the CIMMYT wheat program these progeny are tested first in a series of managed stresses. These are a combination of managed pre-anthesis, post-anthesis and/or continuous stresses generated using limited irrigation in an arid environment (Trethowan et al. 2005) and late planting is used to generate a consistent heat stress from anthesis through the grain-filling period. Selected lines are then tested globally in METs covering the target wheat growing areas of the developing world. The lines selected using these crossing, selection and evaluation principles have performed well globally. Lage and Trethowan (2008) analyzed the performance of synthetic derivatives deployed in the Semi-Arid Wheat Yield Trial distributed by CIMMYT and found that some synthetic derivatives showed superior yield response across a wide range of environments when compared to the best locally adapted cultivars. Synthetic derivatives selected in managed stress environments in Mexico also performed well when tested across variable Australian environments (Ogbonnaya et al. 2007). These authors reported that derivatives yielded 8–30% more than the best locally adapted cultivars, clearly demonstrating a significant genetic correlation between Mexican managed stress environments and sites in Australia.

In a reverse study, Gororo et al. (2002) developed synthetic hexaploid derivatives from locally produced primary synthetics in Australia and tested these materials in both Australia and Mexico. The synthetic derivatives were higher yielding than their elite recurrent parent in 38 of 42 comparisons across environments in both countries, indicating a significant degree of transferability of drought stress response.

There is less information on the response of materials selected under terminal heat stress and tested in high temperature TPEs. In one of the few available studies, Lillemo et al. (2005) analyzed the yield performance of lines distributed globally in CIMMYT's High Temperature Wheat Yield Trial. These lines were developed by late sowing segregating materials in northwestern Mexico with subsequent selection for plant phenotype and grain weight and eventually yield once fixed lines were identified. The selection environment in Mexico clearly correlated with many heat stressed environments globally and materials with stable and superior performance in the heat stress TPE were identified.

In Australia, fixed lines are tested widely in the TPE over a number of years as large genotype x year interactions obscure genetic potential (Chapman et al. 2000). In contrast, in many developing countries fixed lines are tested in METs on research stations, largely for logistical and economic reasons (Ceccarelli and Grando 2007). The materials are only grown on farm once they have been released to farmers. Clearly, the efficiency of cultivar selection from among homozygous lines derived through selection will be dependent on how well the chosen yield testing environments correlate with the TPE.

Nutritional, processing and product quality are generally assessed using grain samples collected from selected genotypes from METs sown across the TPE. The extent of the analysis will reflect the importance of quality within the local, regional and global market place. The analysis of quality is expensive and generally limited to less costly indirect tests in initial assessments, with more detailed analysis of dough properties and end-product quality in subsequent trialing of selected entries.

9.2.3 Some Considerations When Breeding Crops Other than Wheat

We have primarily discussed the adaptation of crop cultivars to a changing climate in the context of wheat. While the principles of assessing and introgressing genetic variability apply across the crop species, regardless of their ploidy level, genome size and geographic distribution, the breeding strategies used to combine this variability will differ among the crop species largely based on their reproductive system. The breeding strategies used to improve the stress tolerance of self-pollinated crops such as wheat, rice and barley are similar, although there are some minor variations. Wheat and barley are more amenable to double haploid production than rice, whereas gene expression is more easily understood in diploids such as barley and rice. There is evidence that wheat hybrids have more stable yield in drier environments (Nehvi et al. 2000), although the cost of producing hybrid seed makes hybrid wheat less attractive than hybrid rice, which is already under cultivation on large areas in Vietnam, India and China (FAO 2005). Nevertheless, the expression of water and heat stress tolerance in hybrid combinations still needs to be assessed. Most hybrid work has focused on the more productive environments where economic returns are greater.

In contrast, significant improvement in the stress response of maize, an open pollinated species, has been achieved through recurrent selection under both prevailing and managed stresses (Bänziger et al. 2004). These schemes increase the frequency of favorable alleles by repeated cycles of selection and intercrossing of superior individuals. Cross specific QTLs (quantitative trait loci) can also be accumulated through recurrent selection using linked molecular markers. In contrast, it is difficult to employ these population improvement techniques in self-pollinated species as it is too expensive to make the required intercrosses among selected progeny. Nevertheless, male sterility (genetic, cytoplasmic and chemically induced) does exist in many self-pollinated species and could be used to facilitate intercrossing and the establishment of recurrent selection schemes.

9.2.4 Conclusion

Plant breeders working in the world's rainfed environments have made steady incremental gains in yield under stress. Over the past 10 years the research investment in plant response to drought and heat has increased significantly, largely driven by improvements in technology, and an increasing awareness of the impending impacts of climate change and reduced water availability on agriculture. Much of this investment has been driven by the private sector in high value crops such as maize (Braun and Brettell 2009). Nevertheless, the investment in wheat and rice, while considerably smaller, has also increased. Improved understanding of the molecular basis of the plant stress response has gone hand-in-hand with improved understanding of the physiological response. International centers such as CIMMYT, the International Center for Agricultural Research in the Drier Areas (ICARDA) and the International Rice Research Institute (IRRI) have extensive breeding and research programs targeting improved water-use and/or tolerance to heat. Other initiatives such as the Generation Challenge Program (GCP) focus on using genetic diversity to improve drought tolerance in crops and provide a mechanism whereby advanced research institutes, national programs in developing countries and international centers such as those mentioned above can bring their skills and resources to bear on these most intractable of problems.

Significant genetic diversity for stress response has already been identified in the primary gene pools of most of the major crop species. However, the challenge of efficiently and effectively introducing this diversity into elite crop backgrounds remains a significant impediment. The vagaries of the production environment, incomplete understanding of the underlying physiological response and the complexity of inheritance of stress responses remain major challenges.

An additional complexity is the relationship between drought and high-temperatures in many production environments. High evapotranspiration rates often lead to increased moisture stress, particularly at lower latitudes. There is significant variation for response to high temperature in most crop gene pools and materials can be selected by simply delaying planting time to expose plants to terminal heat stress. The higher heritability of the selection environment for heat tolerance compared to drought stress should lead to greater gains in productivity under elevated temperature. However, there is evidence that traits important for one stress also influence the other (Table 9.1). Osmotic adjustment, phenology, water-use-efficiency

Table 9.1 Commonality between traits associated with drought and heat stress

Tolerance mechanism	Drought	Heat
Osmotic adjustment	Osmotic adjustment renders plants tolerant to drought stress (Munns 2002; Farooq and Farooq-E-Azam 2001; Blum and Pnuel 1990)	Osmotic adjustment renders plants tolerant to heat stress (Blum and Pnuel 1990)
Rapid growth/earliness	Earliness favors drought tolerance (Blum and Pnuel 1990)	Earliness favors tolerance to heat (Ehlers and Hall 1998; Blum and Pnuel 1990)
Concentration of organic solutes	Increased concentration of organic solutes is observed under drought stress (Sakamoto and Murata 2002; Ashraf and Foolad 2007)	Increased concentration of organic solutes is observed under heat stress (Sakamoto and Murata 2002; Ashraf and Foolad 2007)
Water use efficiency	Water use efficiency of the plant improves drought tolerance (Shannon 1997; Machado and Paulsen 2001; Nultsch 2001)	Water use efficiency of the plant influences heat tolerance (Machado and Paulsen 2001)

and solute concentrations can have an impact on both tolerance to heat and drought. These relationships, if confirmed, will allow the breeder to simultaneously improve both characters.

Increasing levels of CO_2 and higher atmospheric temperatures associated with climate change may at the same time offer both impediments and opportunities (Wahid et al. 2007). A changed climate may favor the cultivation of crops with a C_4 photosynthetic pathway rather than the less efficient C_3 pathway, although all else constant C_3 crops appear to benefit more from increasing levels of CO_2. This has renewed interest in the challenge of converting C_3 crops to the C_4 photosynthetic pathway and a major project is underway at IRRI to produce C_4 rice (http://www.eurekalert.org/pub_releases/2009-01/irri-nhr011909.php).

Clearly genetic diversity is vital to realizing improved crop responses to drought and heat. In some instances there may be insufficient diversity within the crop gene pool to achieve the required levels of improved adaptation. The introduction of transgenes that regulate the plant response to stress may in the future contribute to the overall goal of improved productivity under stress.

References

Ahmad H, Bajelan B (2008) Heritability of drought tolerance in wheat. Am Eurasian J Agric Environ Sci 3:632–635

APA (2004) Population and society: issues, research, policy. In: 12th Biennial Conference, Australian Population Association, Canberra, 15–17 September 2004

Ashraf M, Foolad MR (2007) Roles of glycine betaine and proline in improving plant abiotic stress resistance. Environ Exp Bot 59:206–216

Bänziger M, Setimela PS, Hodson D, Vivek B (2004) Breeding for improved drought tolerance in maize adapted to southern Africa. In: Proceedings of workshop on "Resilient Crops for Water Limited Environments", Cuernavaca, Mexico, 24–28 May 2004

Bertholdsson NO (2005) Early vigour and allelopathy – two useful traits for enhanced barley and wheat competitiveness against weeds. Weed Res (Oxford) 45:94–102

Blum A, Pnuel Y (1990) Physiological attributes associated with drought resistance to wheat cultivars in a Mediterranean Environment. Aust J Agric Res 41:799–810

Blumenthal C, Wrigley CW, Batey IL, Barlow EWR (1994) The heat-shock response relevant to molecular and structural changes in wheat yield and quality. Aust J Plant Physiol 21:901–909

Bonnett DG, Rebetzke GJ, Spielmeyer W (2005) Strategies for efficient implementation of molecular markers in wheat breeding. Mol Breeding 15:75–85

Boubaker M, Yamada T (1991) Screening spring wheat genotypes (*Triticum* sp.) for seedling emergence under optimal and suboptimal temperature conditions. Jpn J Breed 41:381–387

Brakke JP, Francis CA, Nelson LA, Gardner CO (1983) Genotype by cropping system interactions in maize grown in a short season environment. Crop Sci 23:868–870

Braun HJ, Brettell R (2009) The role of international centers in enhancing cooperation in wheat improvement. In: Borlaug Global Rust Initiative, Technical Workshop, Ciudad Obregon, 20 March 2009

Byerlee D, Moya P (1993) Impacts of international wheat breeding research in the developing world, 1966–1990. CIMMYT, Mexico City, Mexico DF

Ceccarelli S, Grando S (2007) Decentralized-participatory plant breeding: an example of demand driven research. Euphytica 155:349–360

Chapman SC, Cooper M, Butler DG, Henzell RG (2000) Genotype by environment interactions affecting grain sorghum. I. Characteristics that confound interpretation of hybrid yield. Aust J Agric Res 51:197–207

Chapman S, Cooper M, Podlich D, Hammer G (2003) Evaluating plant breeding strategies by simulating gene action and dryland environment effects. Agron J 95:99–113

Christensen, JH, Hewitson B, Busuioc A, Chen A, Gao X, Held I, Jones R, Kolli RK, Kwon W-T, Laprise R, Magaña Rueda V, Mearns L, Menéndez CG, Räisänen J, Rinke A, Sarr A, Whetton P (2007) Regional climate projections. In : Solomon S, Qin D, Manning M, Marquis M, Averyt KB, Tignor M, Miller HL, and Chen Z (eds) Climate change 2007: the physical science basis. Contribution of working group I to the fourth assessment report of the intergovernmental panel on climate change. Cambridge University Press, Cambridge, UK and New York, NY, USA

Crossa J, Cornelius PL, Sevedsadr M, Byrne P (1993) A shifted multiplicative model cluster analysis for grouping environments without genotype ranking change. Theor Appl Genet 85:577–586

Crossa J, Burgueno J, Dreisigacker S, Vargas M, Hererra-Foessel SA, Lillemo M, Singh RP, Trethowan R, Warburton M, Franco J, Reynolds M, Crouch JH, Ortiz R (2007) Association analysis of historical bread wheat germplasm using additive genetic covariance of relatives and population structure. Genet 177:1889–1913

DeLacy IH, Basford KE, Cooper M, Bull JK, McLaren CG (1996) Analysis of multi-environment trials, a historical perspective. In: Cooper M, Hammer GL (eds) Plant adaptation and crop improvement. CAB International, Wallingford, UK, pp 39–124

Edmeades G, Banziger M, Campus H, Schussler J (2006) Improving tolerance to abiotic stresses in staple crops. A random or planned process? In: Lamkey KR, Lee M (eds) Plant breeding. The Arnel R. Hallauer International Symposium, Wiley-Blackwell, pp 293–309

Ehdaie B, Whitkus RW, Waines JG (2003) Root biomass, water-use efficiency, and performance of wheat-rye translocations of chromosomes 1 and 2 in spring bread wheat 'Pavon'. Crop Sci 43:710–717

Ehlers JD, Hall AE (1998) Heat tolerance of contrasting cowpea lines in short and long days. Field Crop Res 55:11–21

Elmore RW (1990) Soybean cultivar response to tillage systems and planting date. Agron J 82:69–73

Ezzati M, Lopez AD, Rodgers A, Vander Hoorn S, Murray CJL (2002) Selected major risk factors and global and regional burden of disease. Lancet (British edition) 360:1347–1360

FAO (2001) Climate change: implications for food safety paper. http://www.fao.org/ag/agn/agns/files/HLC1_Climate_Change_and_Food_Safety.pdf
FAO(2005) Rice is life. International year of rice 2004 and its implementation. Food and Agriculture Organization of the United Nations (FAO), Rome, Italy
Farooq S, Farooq-E-Azam (2001) Co-existence of salt and drought tolerance in Triticeae. Hereditas 135(2–3):205
Feldman M (2001) The origin of cultivated wheat. In: Bonjean A, Angus W (eds) The world wheat book. Lavoisier, Paris
Francis CA, Moomaw RS, Rajewski JF, Saeed M (1986) Grain sorghum hybrid interactions with tillage system and planting dates. Crop Sci 26:191–193
Ginkel van M, Ortiz-Monasterio I, Trethowan R, Hernandez E (2001) Methodology for selecting segregating populations for improved N-use efficiency in bread wheat. Euphytica 119:223–230
Gororo NN, Eagles HA, Eastwood RF, Nicolas ME, Flood RG (2002) Use of *Triticum tauschii* to improve yield of wheat in low-yielding environments. Euphytica 123:241–254
Govaerts B, Fuentes M, Mezzalama M, Nicol MJ (2007) Infiltration, soil moisture, root rot and nematode populations after 12 years of different tillage, residue and crop rotation managements. Soil Till Res 94:209–219
Greene SL, Thomas CH, Afonin A (1999) Using geographic information to acquire wild crop germplasm for ex situ collections: II. Post-collection analysis. Crop Sci 39:843–849
Guimaraes EP (2002) Genetic diversity of rice production in Brazil. Pp 11–35. In 'Genetic diversity in rice production: Case studies from Brazil, India and Nigeria". Nguyen VN (ed.), FAO/Rome (Italy). Plant Production and Protection Division
Gupta PK, Kulwal PL, Rustgi S (2005) Wheat cytogenetics in the genomics era and its relevance to breeding. Cytogenet Genome Res 109:315–327
Gutierrez A (2006) Estabilidad del rendimiento y calidad de semilla e industrial de trigos harineros en ambientes de riego y temporal y sistemas de labranza. PhD thesis, Colegio de Postgraduados, Montecillo, Texcoco, Edo de Mexico
Guttieri MJ, Stark JC, O'Brien K, Souza E (2001) Relative sensitivity of Spring wheat grain yield and quality parameters to moisture deficit. Crop Sci 41:335–344
Hede AR, Skovmand B, Reynolds MP, Crossa J, Vilhelmsen AL, Stolen O (1999) Evaluating genetic diversity for heat tolerance traits in Mexican wheat landraces. Genet Resour Crop Evol 46:37–45
Hennessy K, McInnes K, Abbs D, Jones R, Bathols J, Suppiah R, Ricketts J, Rafter T, Collins D, Jones D (2004) Climate Change in New South Wales Part 2: Projected changes in climate extremes. CSIRO website: http://www.dar.csiro.au/publications/hennessy_2004c.pdf
Jiang GL, Shi JR, Ward RW (2007) QTL analysis of resistance to Fusarium head blight in the novel wheat germplasm CJ 9306. I. Resistance to fungal spread. Theor Appl Genet 116:3–13
Joshi AK, Chand R, Arun B, Singh RP, Ortiz R (2007) Breeding crops for reduced-tillage management in the intensive, rice-wheat systems of South Asia. Euphytica 153:135–151
Kennedy G, Nantel G, Shetty P (2002) The scourge of "hidden hunger": global dimensions of micronutrient deficiencies. Food Nutr Agric 32:8–16
Lage J, Trethowan RM (2008) CIMMYT's use of synthetic hexaploid wheat in breeding for adaptation to rainfed environments globally. Aust J Agric Res 59:461–469
Liang YL, Richards RA (1999) Seedling vigor characteristics among Chinese and Australian wheats. Commun Soil Sci Plant Anal 30:159–165
Lillemo M, van Ginkel M, Trethowan RM, Hernandez E, Crossa J (2005) Differential adaptation of CIMMYT bread wheat to global high temperature environments. Crop Sci 45:2443–2453
Mace ES, Xia L, Jordan DR, Halloran K, Parh DK, Huttner E, Wenzl P, Kilian A (2008) DArT markers: diversity analyses and mapping in Sorghum bicolour. BMC Genomics 9:26
Machado S, Paulsen GM (2001) Combined effects of drought and high temperature on water relations of wheat and sorghum. Plant Soil 233:179–187
Melo PGS, Melo LC, Soares AA, Lima LM de, Reis M de S, Juliatti FC, Cornelio VMO (2005) Study of the interaction genotypes × environments in the selection process of upland rice. Crop Breed Appl Biotechnol 5:38–46

Menkir A (2008) Genetic variation for grain mineral content in tropical-adapted maize inbred lines. Food Chem 110:454–464

Mergoum M, Frohberg RC, Stack RW, Rasmussen JB, Friesen TL (2006) Registration of 'Howard' wheat. Crop Sci 46:2072–2073

Moghaddam ME, Trethowan RM, William HM, Rezai A, Arzani A, Mirlohi AF (2005) Assessment of genetic diversity in bread wheat genotypes for tolerance to drought using AFLPs and agronomic traits. Euphytica 141:147–156

Munns R (2002) Comparative physiology of salt and water stress. Plant Cell Environ 25:239–250

Murphy KM, Reeves PG, Jones SS (2008) Relationship between yield and mineral nutrient concentrations in historical and modern spring wheat cultivars. Euphytica 163:381–390

Nehvi FA, Shafiq Wani A, Zargar GH (2000) Heterosis in bread wheat (*Triticum aestivum* L.). Appl Biol Res 2:69–74

Newhouse KE (1985) Genotype by tillage interactions in maize (*Zea mays* L.). Diss Abstr Int B (Sci Eng) 45:1973B

Nultsch W (2001) Allgemeine Botanik. 11. Auflage. Georg Thieme Verlag, Stuttgart

Ogbonnaya F, Dreccer F, Ye G, Trethowan RM, Lush D, Shepperd J, Ginkel M van (2007) Yield of synthetic backcross-derived lines in rainfed environments of Australia. Euphytica 157:321–336

Okogbenin E, Porto MCM, Egesi C, Mba C, Espinosa E, Santos LG, Ospina C, Marín J, Barrera E, Gutiérrez J, Ekanayake I, Iglesias C, Fregene MA (2007) Marker-assisted introgression of resistance to cassava mosaic disease into latin American germplasm for the genetic improvement of cassava in Africa. Crop Sci 47:1895–1904

Ozturk L, Yazici MA, Yucel C, Torun A, Cekic C, Bagci A, Ozkan H, Braun HJ, Savers Z, Cakmak I (2006) Concentration and localization of zinc during seed development and germination in wheat. Physiol Plant 128:144–152

Pingali P (2007) Westernization of Asian diets and the transformation of food systems: implications for research and policy. Food Policy 32:281–298

Rajaram S, Mann CE, Ortiz-Ferrara G, Mujeeb-Kazi A (1983) Adaptation, stability and high yield potential of certain 1B/1R CIMMYT wheats. In: Sakamoto S (ed) Proceedings of 6th International Wheat Genetics Symposium. Maruzen, Kyoto, pp 613–621

Rebetzke GJ, Richards RA, Sirault XRR, Morrison AD (2004) Genetic analysis of coleoptile length and diameter in wheat. Aust J Agric Res 55:733–743

Rebetzke GJ, Richards RA, Fettell NA, Long M, Condon AG, Forrester RI, Botwright TL (2007) Genotypic increases in coleoptile length improves stand establishment, vigour and grain yield of deep sown wheat. Field Crop Res 100:10–23

Reddy BVS, Ramesh S, Longvah T (2005) Prospects of breeding for micronutrients and beta-carotene-dense sorghums. Int Sorghum Millets Newsl 46:10–14

Reynolds MP, Trethowan RM (2007) Physiological interventions in breeding for adaptation to abiotic stress. pp 129–146. In: Spiertz JHJ, Struik PC, Van Laar HH (eds) Scale and complexity in plant systems research, gene-plant-crop relations. Springer, The Netherlands

Reynolds M, Dreccer F, Trethowan RM (2007b) Drought adaptive mechanisms from wheat landraces and wild relatives. J Exp Bot 58:177–186

Reynolds MP, Pierre CS, Saad ASI, Vargas M, Condon AG (2007a) Evaluating potential genetic gains in wheat associated with stress-adaptive trait expression in elite genetic resources under drought and heat stress. Crop Sci 47(Suppl 3):S172–S189

Ribaut JM, Hoisington DA, Deutsch JA et al (1996) Identification of quantitative trait loci under drought conditions in tropical maize. 1. Flowering parameters and the anthesis-silking interval. Theor Appl Genet 92:905–914

Sakamoto A, Murata N (2002) The role of glycine betaine in the protection of plants from stress: clues from transgenic plants. Plant Cell Environ 25:163–171

Sears ER (1976) Genetic control of chromosome pairing in wheat. Annu Rev Genet 10:31–51

Shannon MC (1997) Adaptation of plants to salinity. Adv Agron 60:75–120

Sharma DL, Anderson WK (2004) Small grain screenings in wheat: interactions of cultivars with season, site, and management practices. Aust J Agric Res 55:797–809

Skovmand B, Reynolds MP, DeLacy IH (2001) Mining wheat germplasm collections for yield enhancing traits. Euphytica 119:25–32

Skylas DJ, Cordwell SJ, Hains PG, Larsen MR, Basseal DJ, Walsh BJ, Blumenthal C, Rathmell W, Copeland L, Wrigley CW (2002) Heat shock of wheat during grain filling: proteins associated with heat-tolerance. J Cereal Sci 35:175–188

Trethowan RM (2007) Breeding wheat for high iron and zinc at CIMMYT: state of the art, challenges and future prospects. In: Proceedings of Seventh International Wheat Conference. Mar del Plata, Argentina

Trethowan RM, Mujeeb-Kazi A (2008) Novel germplasm resources for improving environmental stress tolerance of hexaploid wheat. Crop Sci 48:1255–1265

Trethowan RM, Singh RP, Huerta-Espino J, Crossa J, Ginkel M van (2001a) Coleoptile length variation of near-isogenic *Rht* lines of modern CIMMYT bread and durum wheats. Field Crop Res 70:167–176

Trethowan RM, Crossa J, Ginkel M van Rajaram S (2001b) Relationships among bread wheat international yield testing locations in dry areas. Crop Sci 41:1461–1469

Trethowan RM, Reynolds M, Sayre K, Ortiz-Monasterio I (2005) Adapting wheat cultivars to resource conserving farming practices and human nutritional needs. Ann Appl Biol 146:405–413

Trethowan RM, Manes Y, Chattha T (2009) Breeding for improved adaptation to conservation agriculture improves crop yields. Proceedings of the 4th International Congress on Conservation Agriculture, New Delhi, February 4–7 NASC Complex, Pusa, New Delhi 110 012 (in press)

Uauy C, Distelfeld A, Fahima T, Blechl A, Dubcovsky J (2006) A NAC gene regulating senescence improves grain protein, zinc, and iron content in wheat. Science 314:1298–1301

Ullrich SE, Muir CE (1986) Genotypic response of spring barley to alternative tillage systems. Cereal Res Commun 14:161–168

Villareal RL, Mujeeb-Kazi A (1999) Exploiting synthetic hexaploids for abiotic stress tolerance in wheat. In: Proceedings of the tenth regional wheat workshop for Eastern, Central and Southern Africa, University of Stellenbosch, South Africa, 14–18 September 1998, pp 542–552

Villareal RL, del Toro E, Mujeeb-Kazi A, Rajaram S (1995) The 1BL/1RS chromosome translocation effect on yield characteristics in a *Triticum aestivum* L. cross. Plant Breeding 114:497–500

Wahid A, Gelani S, Ashraf M, Foolad MR (2007) Heat tolerance in plants: an overview. Environ Exp Bot 61:199–223

Warburton ML, Crossa J, Franco J, Kazi M, Trethowan R, Rajaram S, Pfeiffer W, Zhang P, Dreisigacker S, van Ginkel M (2006) Bringing wild relatives back into the family: recovering genetic diversity of CIMMYT bread wheat germplasm. Euphytica 149:289–301

Weightman RM, Millar S, Alava J, Foulkes MJ, Fish L, Snape JW (2008) Effects of drought and the presence of the 1BL/1RS translocation on grain vitreosity, hardness and protein content in winter wheat. J Cereal Sci 47:457–468

Welch RM, Graham RD (2004) Breeding for micronutrients in staple food crops from a human nutrition perspective. J Exp Bot 55:353–364

William HM, Trethowan R, Crosby-Galvan EM (2007) Wheat breeding assisted by markers: CIMMYT's experience. Euphytica 157:307–319

Yang J, Sears RG, Gill BS, Paulsen GM (2002) Growth and senescence characteristics associated with tolerance of wheat-alien amphiploids to high temperature under controlled conditions. Euphytica 126:185–193

Part IV

Chapter 10
Global and Regional Assessments

David Lobell and Marshall Burke

Abstract The main conclusions from recent global and regional assessments are reviewed, with an emphasis on China, India, Africa, and the United States. Most studies have provided primarily "best-guess" point estimates, often supplemented with a few sensitivity analyses, but without a comprehensive measure of uncertainties. Although some useful lessons have been learned, most existing estimates of food security risks leave much to be desired. We explore these estimates, some of their strengths and weaknesses, and some additional opportunities for measuring uncertainties.

10.1 Introduction

This book has focused on the theory and data behind models used to evaluate climate change impacts, rather than on the output of such models. In part this was because knowing the output of models is of little help without an understanding of the capabilities and limitations of the underlying models, and in part because the current pace of new research on applications of models is so rapid that a book devoted to this topic is sure to be outdated in a few years. Yet we recognize that a current summary of the literature will be useful to most readers, and so here provide a brief review of recent findings. For more exhaustive reviews, the readers are encouraged to consult recent assessments by various international groups (e.g., Easterling et al. 2007).

As with any discussion of impacts, a useful starting point is to define the scale and variable of interest. As described in Chapter 2, most people in the world are net buyers of food, and consume diets dominated by the three main staples (rice, wheat, and maize). The prices of these commodities are therefore among the most relevant to food security. Most people also live in communities that trade beyond their local borders, in many cases with places across the globe. For this reason the price of food in given country often depends more on global supply and demand than on local production.

A global perspective is therefore critical to assessing climate change impacts, even if one is interested in a single country (Reilly et al. 1994). Yet assessments for

D. Lobell (✉) and M. Burke
Stanford University, CA, USA

D. Lobell and M. Burke (eds.), *Climate Change and Food Security*,
Advances in Global Change Research 37, DOI 10.1007/978-90-481-2953-9_10,

individual regions or countries can still be useful, in two main ways. First, they allow more detailed study of local climate and crop changes that can feed into global assessments, which often have very rough estimates of regional yield responses.[1] Second, they can focus on local adaptation options that would improve local yields in the face of climate change, regardless of global price changes. This chapter therefore addresses both global and regional studies.

10.2 Global Assessments

The first major studies of global agricultural impacts began roughly 20 years ago, as agriculture was one of the first sectors for which impacts of climate change were thought to be important (Kane et al. 1992; Rosenzweig and Iglesias 1994; Rosenzweig and Parry 1994). Then, as now, these efforts focused on linking three basic modeling pieces that had been previously developed and applied independently: (1) models of climate response to higher CO_2; (2) models of crop yield responses to climate change, higher CO_2, and, in some cases, potential farmer adaptations; and (3) models of adjustments in the world food economy in response to differential yield effects in different regions (Fig. 10.1). Some studies focused solely on changes in worldwide aggregate production and prices, while others also investigated changes in food security, typically measured by the number of malnourished.

One of the first seminal assessments considered the global impacts of doubling CO_2 from pre-industrial levels (Rosenzweig and Parry 1994) by utilizing a network of crop modelers from around the world, who provided estimates of yield impacts from locally calibrated models for prescribed climate changes in over 100 sites in 18 countries. These site-level estimates were then used to infer national level

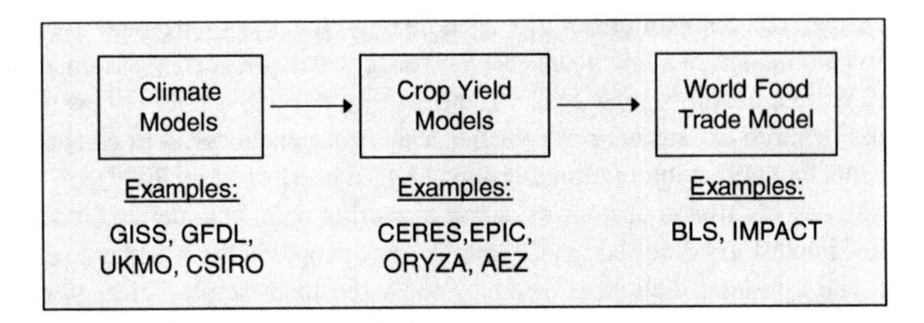

Fig. 10.1 Outline of a common approach to estimating global impacts of climate change on food prices, production, trade, and hunger. Climate scenarios generated from climate models are used to drive crop yield models for individual locations. The simulated yield responses are then summed for different regions and input into a food trade model, which determines the equilibrium price for each commodity and the associated crop areas, yields, production, and trade for each region. The price changes are also often used to estimate the change in number of people at risk of hunger

[1] For example, yield changes in Rosenzweig and Parry (1994) were prescribed by interpolating results from crop model simulations at individual sites, of which the only sites in sub-Saharan Africa were for maize in Zimbabwe.

production changes for all cereals in all countries, based on similarities of agronomic characteristics among crops and agro-ecological environments among countries. The results were then aggregated into regional yield changes according to the regions defined in the Basic Linked System (BLS) model of agricultural trade.

Figure 10.2 presents the yield changes used as input to BLS for the four major commodity groups treated in that study: wheat, rice, coarse grains, and oilseeds. These

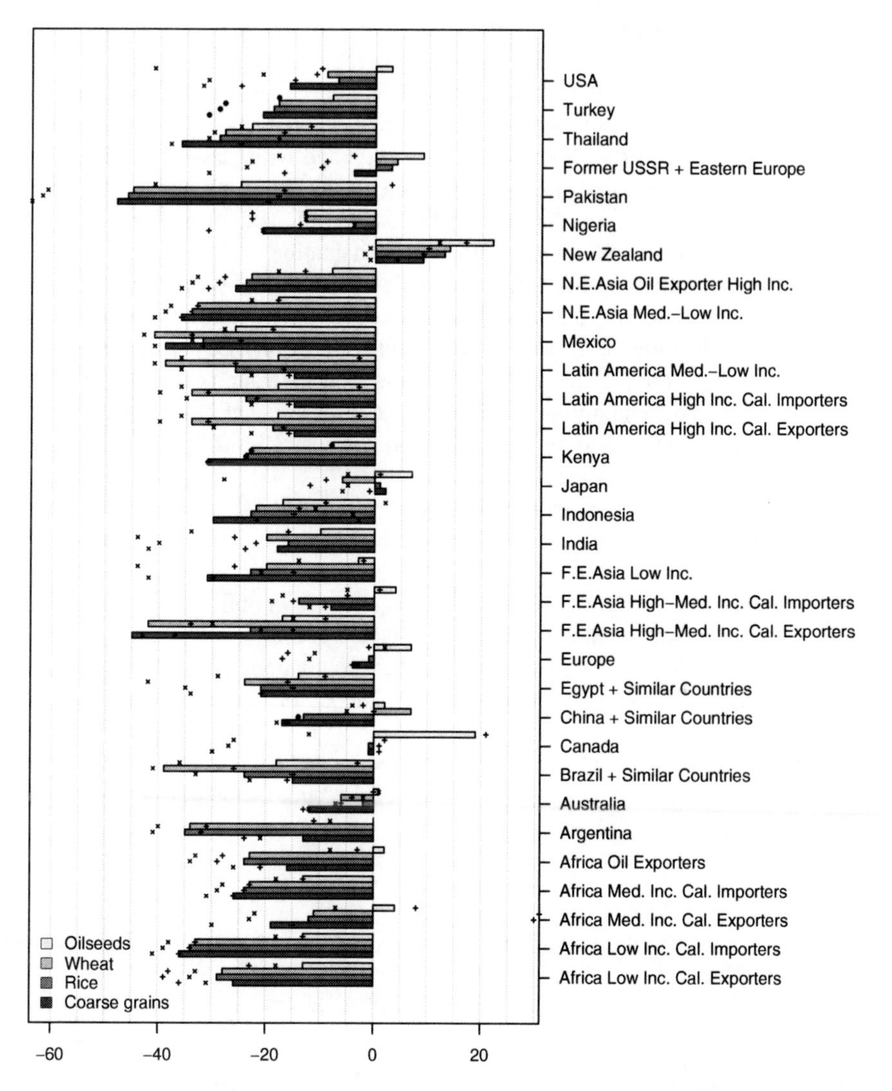

Fig. 10.2 Yield changes for doubled CO_2 (climate change plus CO_2 fertilization effects) used as input into global trade model in Rosenzweig and Parry (1994), based on crop model simulations for 100+ sites. *Bars* show yield changes using climate scenarios from the NASA GISS climate model, "+" indicates values for GFDL climate model, and "x" for UKMO climate model. The magnitude of CO_2 fertilization was 4%, 11%, 12%, and 17% for coarse grains, rice, wheat, and oilseeds, respectively. All yield changes correspond to simulations without any farmer adaptation

results exhibit some of the main features of yield changes in most global assessments. First, the net yield impacts of climate change and doubled CO_2 tends to be negative for most but not all region–commodity combinations. Second, high latitude countries tend to have lower impacts than tropical countries because they start from a cooler baseline. Third, C_3 crops (rice, wheat, and oilseeds) tend to have lower impacts than maize because of greater CO_2 fertilization. Fourth, yield impacts varied substantially for different climate model scenarios, three of which were considered in the study.

An important innovation by Rosenzweig and colleagues was to examine the potential impact of adaptation in a systematic way. Each modeling group was asked to perform simulations with no adaptation (i.e. climate change and CO_2 effects only, as shown in Fig. 10.2), with "level 1" adaptations, where tactical decisions such as planting date and cultivar choice were optimized, and with "level 2" adaptations, which included more costly adaptations such as development of new irrigation infrastructure and new crop varieties. This design allowed an evaluation of the benefits of both small and large investments in adaptation.

Table 10.1 summarizes the global production changes that result from the yield changes illustrated in Fig. 10.2, as well as those under different adaptation scenarios. The authors concluded that, assuming the full effects of CO_2 fertilization were realized, impacts on global cereal production ranged from negligible to slight declines (<10%) depending on the climate model used. Level 1 adaptations had a fairly small effect on overall impacts, but more expensive level 2 adaptations were effective in minimizing negative outcomes.

The associated changes in cereal prices for doubled CO_2 and no adaptation ranged from 25% to 150% for the three climate scenarios, with increases in the number of malnourished by 10–60% (malnourishment prevalence in the BLS model increased by roughly 1% for each 2.5% increase in prices). For adaptation level 2, when global production changes ranged from +1 to −2%, price changes ranged from −5% to +35%, and malnourished populations changed by between −2% and +20%. The role of on-farm vs trade adaptations in these projections are discussed further in Chapter 8.

Many subsequent global assessments have been conducted since the early 1990s (Reilly et al. 1994; Parry et al. 1999; Fischer et al. 2002; Darwin 2004; Parry et al. 2004; Fischer et al. 2005), providing some consensus on several key points:

Table 10.1 The projected impacts of doubled CO_2 on global cereal production (% change) for different climate models, adaptation levels, and with and without CO_2 fertilization (from Rosenzweig and Parry 1994; see text for details on adaptation levels)

Scenario	GISS	GFDL	UKMO
Climate change only	−11	−12	−20
With CO_2 fertilization	−1	−3	−8
With CO_2 and adaptation Level 1	0	−2	−6
With CO_2 and adaptation Level 2	1	0	−2

Climate models: GISS = Goddard Institute for Space Studies (4.2, 11); GFDL = Geophysical Fluid Dynamics Laboratory (4.0, 8); UKMO = United Kingdom Meteorological Office (5.2, 15). Numbers in parentheses are global average change in temperature (°C) and precipitation (%) for each model.

(i) Global price increases associated with a doubling of CO_2 range from negligible for moderate climate change to significant for more extreme climate scenarios. As a doubling of CO_2 is likely to be reached near mid-century, studies that evaluate transient scenarios out to 2100 generally also consider higher CO_2 levels. Long-term price impacts tend to increase as CO_2 levels are increased, because the positive effects of higher CO_2 are increasingly outweighed by the negative impacts of associated climate changes.[2] Equilibrium price changes thus often exceed 10% by the end of the century for most emissions scenarios (Easterling et al. 2007).

(ii) Impacts are generally more negative for developing countries than developed countries. This arises mainly from the fact that most developing countries are in tropical climates with a warmer baseline climate, so that warming more quickly pushes crops beyond their optimum temperature range. In addition, tropical countries tend to rely more on C_4 crops like maize, sorghum, and millet that exhibit small CO_2 fertilization effects. As a result of more detrimental effects in developing nations, trade models anticipate substantial expansion of trade flows from North to South.

(iii) Although the general North-South gradient in impacts is seen in most models, there can be substantial heterogeneity within regions owing to the specific patterns of rainfall and temperature changes. For example, the United States exhibited among the most severe yield losses out of all nations in a study that used the Hadley Center's HadCM3 model (Parry et al. 1999). Even neighboring countries can exhibit quite different responses depending on details of rainfall simulations. This is illustrated by simulated cereal yield impacts by 2050 in India and Pakistan from Fischer et al. (2002), where impacts ranged from 16% higher to 10% lower in India relative to Pakistan depending on the climate model (Fig. 10.3).

(iv) Adaptations can substantially reduce the impacts of climate change, but relatively easy options such as planting date shifts generally have only a small impact while more expensive changes provide most of the benefit (see Chapter 8). The simulated benefits of adaptations are tempered by two caveats. First, few studies have explicitly incorporated the costs of these adaptations into measures of economic impact, nor have they performed a clear cost-benefit analysis. Second, most studies find that adaptation is likely to proceed more effectively in developed nations, thus exacerbating the North–South gradient in impacts and the trade imbalances that result. Additional issues such as how quickly farmers can actually perceive climate trends are discussed in Chapter 8.

10.3 Regional Assessments

In addition to estimates of regional yield changes that have been developed in the process of global assessments (e.g., Fig. 10.2), there is also a growing wealth of studies focused on particular regions. Indeed, the literature is too vast to provide an exhaustive review.

---

[2] However, higher emissions scenarios can actually reduce near-term impacts since the CO_2 fertilization effect responds instantly to higher CO_2 levels while the climate system takes several decades to respond.

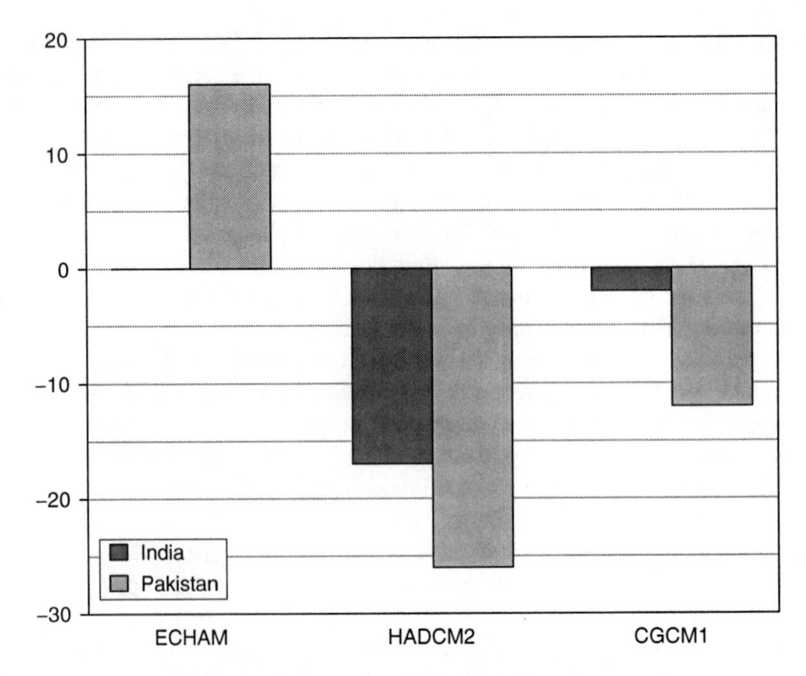

Fig. 10.3 Yield changes in India and Pakistan by 2050 (relative to 1961–1990) simulated with the Agro-ecological Zone (AEZ) models of Fischer et al. (2002) for three different climate model scenarios: ECHAM (Max-Planck Institute of Meteorology), HADCM2 (Hadley Centre for Climate Prediction and Research), and CGCM1 (Canadian Centre for Climate Modelling and Analysis). Not only do average yield changes vary with climate scenario, but the relative impacts in India and Pakistan differ greatly, likely due to the spatial distribution of rainfall change

Instead we present below a brief summary for four key regions. We focus here on projected impacts in the absence of adaptation, with potential effects of adaptation addressed in Chapter 8.

10.3.1 China

Rice remains the main staple in China as it has for thousands of years, accounting for over one-quarter of the calories and roughly one-sixth of the protein consumed in 2003 (FAO 2007). Wheat, soybean, and maize are also important components of the modern Chinese diet, consumed both directly and indirectly via animal products (Chapter 2). Nearly all rice fields in China are irrigated (Huke and Huke 1997), while the majority of wheat and maize fields are rainfed.

China is commonly viewed as facing relatively benign impacts of climate change on agriculture. For example, Rosenzweig and Parry (1994) projected moderate yield declines for rice and maize but slight increases for wheat and soybean (Fig. 10.2). A main reason for the modest impacts is that most of the rainfed crops are concentrated in Northern China where fairly cool temperatures predominate for most of the growing season. In Southern China, where rice is the dominant crop, widespread irrigation is assumed to prevent any significant losses that would arise from greater water stress.

Even with these moderating factors, warming is expected to harm yields in most assessments. Figure 10.4a summarizes estimates of rice yield changes from various crop-modeling studies of China that considered a range of warming. Studies often differ by a factor of two, depending on the crop model used, the projected change in rainfall, and other factors. Yet all show a negative response for warming. A recent Ricardian analysis of revenues from 8,405 households throughout China also found a negative marginal impact of temperature on the average crop revenues (Wang et al. 2008).

Thus, most projected gains in agriculture in China result from a fertilization effect of CO_2 that is simulated to overwhelm climate related losses. This CO_2 effect is illustrated for rice in Fig. 10.4 by the arrows, whose lengths indicate that the magnitude of CO_2 fertilization also differs considerably by study. Some prominent studies appear to include CO_2 effects that are much bigger than suggested by recent field experiments (see Chapter 7). For example, one analysis of impacts by 2050 projected rice, wheat, and maize yield losses under an A2 emission scenario of 12.4, 20.4, and 22.8%, respectively, without CO_2 fertilization, but yield gains of 6.2, 20.0, and 18.4% with CO_2 fertilization (Lin et al. 2005; Xiong et al. 2007). This corresponds to 18.6, 40.4, and 41.2% yield boost from CO_2 concentrations of 559 ppm, reflecting a major role of reduced water stress from stomatal closure in the version of the CERES models used in that study.

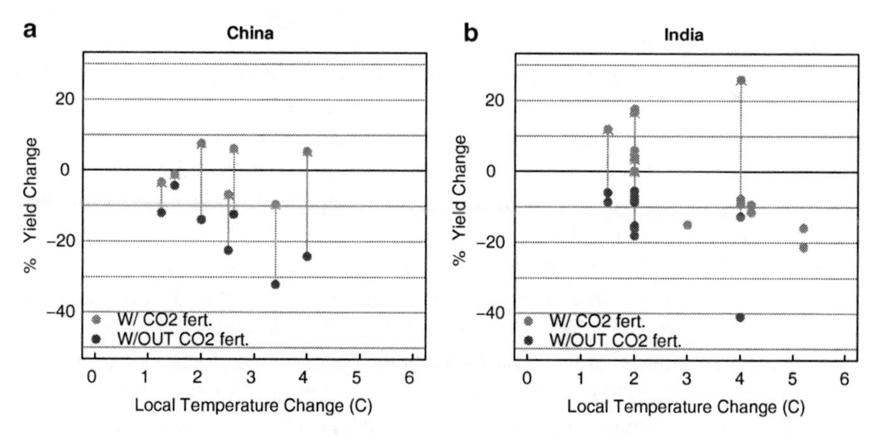

Fig. 10.4 Crop model estimates of rice yield changes for different levels of warming for (**a**) China and (**b**) India, as reported in various studies. *Black dots* indicate effects without CO_2 fertilization, and *gray dots* with CO_2 fertilization, with *arrows* connecting points from the same study. The only difference between points connected by arrows is thus the simulated effect of CO_2. Values were derived from three studies for China (Matthews 1995; Lin et al. 2005; Tao et al. 2008), and five for India (Matthews 1995; Lal et al. 1998; Saseendran et al. 2000; Aggarwal and Mall 2002; Krishnan et al. 2007)

Most yield impact assessments for China to date assume a constant supply of irrigation water. Given the crucial role that irrigation plays in Chinese agriculture, however, the potential of climate change to increase or decrease water availability and demand is also of concern. Tao et al. (2008) estimated water use in rice at five stations in China for various climate scenarios, and found in nearly all cases that a shortened growing season resulted in overall lower crop evapotranspiration and water use. Water use was further curtailed when CO_2 effects on stomatal closure were taken into account, even when yield gains were simulated. As a result of lower water use, irrigation demand was reduced in nearly all cases except where precipitation was projected to decline. These results suggest that shorter seasons and lower ET rates will tend to diminish water use in agriculture, but the demand for (and surely the availability of) irrigation water will also depend on uncertain precipitation trends.

10.3.2 India

Indian agriculture is characterized by a wide range of crops, most prominent among them rice (44 Mha harvested in 2007), wheat (28 Mha), and millet (11 Mha) (FAO 2007). In contrast to China, most crops in India are grown in relatively hot conditions. Spring and summer temperatures commonly exceed 40°C even in the current climate. Thus, one would expect crops to be more sensitive to warming. Indeed, a survey of rice crop modeling studies indicate that even with CO_2 fertilization, warming above 2°C is likely to lower Indian rice yields (Fig. 10.4b). Only a single study with nearly a 40% boost from higher CO_2 (Matthews 1995) shows yield gains for more than 2°C warming.

The sensitivity of Indian crops to warming is also evident in statistical analysis of time series data (Table 10.2). Using the approach outlined in Chapter 5 to estimate impacts by 2030, many important crops are anticipated to incur yield losses (Lobell et al. 2008). Combined with the fact that India possesses the highest population of undernourished people in the world (Chapter 2), food security in India appears particularly vulnerable to climate change.

In a Ricardian study, net farm revenues in India were similarly found to respond negatively to warming, with a 12% reduction in revenue for a scenario with 2°C warming and a 7% increase in rainfall (Dinar et al. 1998). Thus, whether using crop models, time series based models, or panel based methods, the expected effects of warming are negative for most crops. In the near term, CO_2 benefits may counteract these losses, although not in the case of prominent C_4 crops such as millet and sugarcane.

Like China, India is heavily reliant on irrigation and thus the future reliability of water resources will likely play a crucial role in determining the net impact of climate change. Declines in irrigation water, whether resulting from climate change or other factors such as increased urban demand, could greatly increase the sensitivity of crops to higher temperatures. For example, crop model simulations for wheat in Northwest India suggest that the net impact of 1°C warming is roughly double

Table 10.2 Estimated sensitivity of average national yields of Indian crops to a 1°C rise in average growing season temperature, based on time series analysis of 1961–2002 data (adapted from Lobell et al. 2008). The model included both average growing season temperature and rainfall. In several cases (e.g., millet) most of the model's predictive skill came from rainfall, while temperature sensitivities were not significant

			Inferred yield change (%) per °C	
Crop	Percentage contribution to calories in Indian diet	Model r^2	Mean	Standard deviation
Wheat	22	0.27	−2.6	0.7
Rice	27	0.63	−4.0	2.0
Sugarcane	14	0.03	−0.1	2.2
Millet	3	0.63	−4.2	4.4
Sorghum	2	0.14	0.8	6.5
Maize	2	0.16	−3.6	2.5
Soybean	2	0.11	−7.4	5.5
Groundnut	2	0.67	−3.4	5.5
Rapeseed	2	0.45	−7.4	2.5

for a scenario reflecting severe water conservation than under business as usual irrigation (Lal et al. 1998). Unfortunately, the future effects of climate change on water resources in South Asia have not, to our knowledge, been closely examined as of this writing.

10.3.3 Sub-Saharan Africa

Each of the 47 countries in sub-Saharan Africa (SSA) has its own mix of primary crops and diets (Chapter 2). Nonetheless, as a whole Africa can be characterized as a continent heavily dependent on C_4 cereals (maize, sorghum, and millets), cassava, groundnuts, and, to a lesser extent, rice and wheat. SSA is also generally characterized by hot growing conditions relative to much of the developed world, although Southern African growing seasons can be relatively cool. Outside of a few countries, such as Sudan, irrigation is very rare in SSA, with roughly 5% of cereal area in irrigation as of 1995 and little growth expected by 2025 (Rosegrant et al. 2002).

Given these conditions – a warm baseline climate, a lack of irrigation, and a predominance of C_4 crops unlikely to respond strongly to higher CO_2 – it is not surprising that model projections of climate impacts on SSA crops have tended to be negative. In a study using CERES-Maize for all countries in SSA and Latin America, Jones and Thornton (2003) project a fairly average modest decline of 10% by 2055 using a climate scenario from the Hadley CM2 model, though they point to a wide range of impacts between and even within several countries.

Others have suggested more negative impacts, most notably the fourth assessment report of the IPCC, whose chapter on Africa concludes that "reductions in yield in some countries could be as much as 50% by 2020" (Boko et al. 2007).

Though this statement is accompanied only by a citation of a discussion paper on historical losses in drought years in Northern Africa, it nonetheless received widespread media attention.

There has been less work specifically devoted to SSA than many other regions, and as a result the understanding of potential outcomes and uncertainties has been quite limited (Challinor et al. 2007). The World Bank recently commissioned a series of cross-sectional studies of crop revenue in Africa (e.g., Kurukulasuriya et al. 2006), with mixed results but generally negative impacts. Time series models indicate that Southern Africa is extremely sensitive to warming, more so than the warmer tropical regions (Lobell et al. 2008). A likely reason for this is that fertilizer rates and average yields in Southern Africa are considerably higher, so that there is more room for damage in hot years relative to cool years.

Combining data from all countries into a panel analysis for the 1961–2002 period, a recent analysis by Schlenker and Lobell (2009) attempted to estimate the probability of different levels of yield impacts by 2050 for five major crops in SSA: maize, sorghum, millet, groundnuts, and cassava. The first four were found to have significant negative responses (not accounting for CO_2 fertilization), even in the case of millet that is generally regarded as relatively tolerant of hot conditions. There was no clear relationship between cassava production and either temperature or rainfall, likely because cassava harvests are irregular and therefore collection of production data and definition of growing season weather are much more difficult than for other crops.

A comparison of the impact probability distributions for maize with point-estimates from previous studies is shown in Fig. 10.5 for four countries (for a more complete comparison, see Schlenker and Lobell 2009). The estimates of Parry et al. (1999) and Jones and Thornton (2003) generally fall within the distributions, although they tend toward the optimistic end of the range in most countries. Impacts projected by the FAO model (Fischer et al. 2002) in contrast appear much more optimistic than the other three studies. Since Fischer et al. (2002) only report estimates

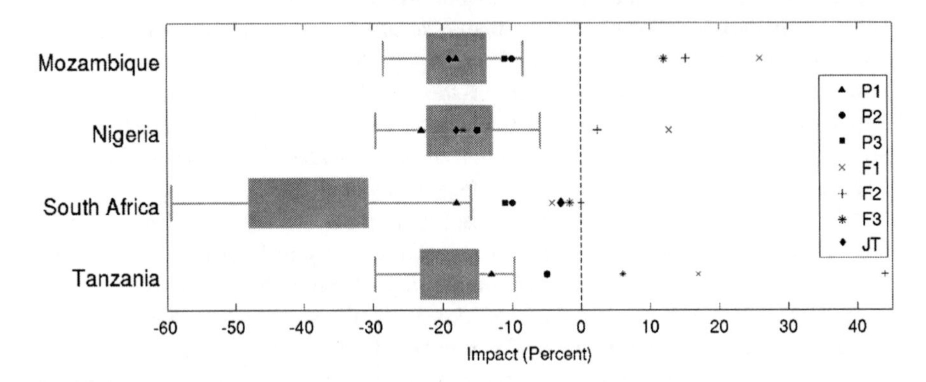

Fig. 10.5 Estimated probability distribution of maize yield impacts of climate change, without adaptation, in selected African countries, based on Schlenker and Lobell (2009). *Gray bars* show 25th–75th percentile of estimates, *whiskers* show 5th–95th percentile, and *middle vertical line* shows median projection. Point estimates from Parry et al. (1999) ("P"), Fischer et al. 2002 ("F"), and Jones and Thornton (2003) ("JT") are shown for comparison

of impacts with CO_2 fertilization, we have subtracted 4% from each projection, equivalent to the reported CO_2 response for the AEZ model (Tubiello et al. 2007). This may explain some of the positive bias in the FAO relative to the other models, but in general the FAO results appear hard to justify, especially given the limited documentation of the model's performance in simulating present-day African yields. Overall, the results in Fig. 10.5 support the notion that climate change presents a serious risk to African crop yields, and that losses by 2050 could easily exceed 20% in many countries.

10.3.4 United States of America

The United States dominates international trade of several agricultural commodities. It is the leading exporter of maize, soybeans, and wheat flour, and therefore production of these crops in the United States has an important influence on food prices throughout the world and thus on food security. The United States is also one of the most extensively studied regions in terms of climate change impacts. An early analysis based on crop models indicated that some Northern regions would gain from temperature increases, but that in most important production regions, such as the Corn Belt and Southern Great Plains, warming accompanying a doubling of CO_2 would reduce yields by an amount roughly equal to the fertilization effect of CO_2 (Adams et al. 1990). Thus, it was concluded that a doubling of CO_2 would result in small net changes to major commodities in the United States, and that adaptation to warming could even result in net benefits.

These conclusions have been generally supported by many subsequent studies, many of which are reviewed in a recent national summary report (CCSP 2008). This report also highlights some factors that have not been considered in most modeling studies, such as the likely northward migration of weeds. The range of several important pests, such as corn earworm, are also currently constrained by winter temperatures and can be expected to expand greatly in future climates (Diffenbaugh et al. 2008). There may also be smaller than expected CO_2 benefits for the most widely grown crop, maize (see Chapter 7). Finally, there might also be greater impacts from extreme heat episodes than most currently used models anticipate. For instance, Schlenker and Roberts (2008) find that yields of maize, soybeans, and cotton in the United States are very sensitive to extreme heat (see Chapter 6). Such results raise important questions about the generally low sensitivity of process-based crop models to warming in these systems.

As some key regions such as California and Nebraska are heavily dependent on irrigation, changes in water resources will also be important. For much of the country, the future direction of rainfall trends remains ambiguous and therefore water availability may increase or decrease (Thomson et al. 2005a). In the West, heavy dependence on snowpack combined with anticipated higher temperatures will very likely result in reduced water availability during the growing season months (Maurer and Duffy 2005). Interestingly, total national irrigated area was projected to decrease across a range of climate scenarios, because rainfed crops became more competitive in scenarios of rainfall increases while irrigation water became limiting in scenarios of drying (Thomson et al. 2005b).

10.4 Measuring Uncertainties

The conclusions outlined above represent, in most cases, a relatively broad consensus among researchers. However, it is important to emphasize that all global and most regional assessments to date can best be characterized as "best-guess," usually supplemented with some simple sensitivity tests such as impacts with and without CO_2 fertilization or adaptation. Yet a consensus among best guesses does not imply that we fully understand the risks associated with climate change, even at global scales (see Chapter 1). True uncertainty analysis, which attempts to quantify the total uncertainty inherited from all individual sources of uncertainty and estimate probabilities of different outcomes, has generally been absent. In some cases, failing to consider uncertainties can lead to a false sense of confidence about projections. In other cases, simple sensitivity tests can overstate uncertainties, such as comparing results from models run with and without CO_2 fertilization (i.e. 30% or 0% fertilization) when the true range of uncertainty likely lies in between.

In our opinion, the remaining key need in impact assessments is to better characterize uncertainty and risks. This may include incorporating new processes (such as ozone or pest damage) but will likely center on better understanding the processes already treated in current models. Accomplishing this task is likely beyond the means of any single research group, given that no group has the number or diversity of models needed to evaluate the full suite of uncertainty sources. A proven strategy for assessing uncertainty is thus to compare model outputs from different groups, so-called model intercomparison projects (MIPs), such as is commonly done with climate models (see Chapter 3).

Periodic literature reviews and syntheses, such as those by the IPCC, provide useful insight into uncertainties but are not true MIPs in two important respects. First, studies often differ widely in the specific questions they address, and as a result typically evaluate different outcomes, time scales, and spatial scales of interest. For example, many studies have used equilibrium scenarios of doubled CO_2, while others have used transient climate change projections. In the former, the climate system has come to equilibrium with atmospheric CO_2 and therefore tends to be warmer than climate at the time of CO_2 doubling in a transient simulation, because the climate system takes decades to respond to changes in CO_2. Doubled CO_2 experiments are therefore difficult to interpret as projections for any particular year. A more straightforward disparity is that many studies examine only production impacts or global commodity prices while others calculate changes in number of malnourished.

A second major challenge in the absence of MIPs is that most studies do not examine all relevant sources of uncertainty, and even a collection of studies will often all treat some model components in the same way. For example, most existing global assessments employ the same economic trade model (BLS), so that uncertainties associated with the structure of the trade model cannot be fruitfully explored.

Only when multiple groups follow the same model experiment design can a sizable number of simulations for the same variable of interest, and a systematic evaluation of the main potential sources of uncertainty, be ensured. One of the main obstacles to implementing MIPs is the substantial amount of foresight, coordination, and resources required to run multiple combinations of state-of-the art climate, crop, and trade models. But such MIPs will be crucial if we are to make progress in measuring uncertainties in impact or adaptation assessments.

An alternative, although by no means a substitute, to MIPs is to represent uncertainty in each component with simple statistical models that are computationally much more efficient and can readily represent uncertainties. This approach is exemplified by Tebaldi and Lobell (2008), who attempted to compute probability distributions of climate change impacts on global average yield changes for maize, wheat, and barley production for 2030 (see Chapter 3).

With all of the potential sources of uncertainty, one may wonder whether it is really necessary to examine each equation in each model or if a few key equations deserve most of the scrutiny. In fact, evaluating each equation is likely neither feasible nor necessary, but which should we focus on? Some insight into this question can be gained by evaluating projections with individual factors varied one at a time over a plausible range of values. This type of sensitivity analysis was used by Lobell and Burke (2008) to evaluate sources of uncertainty for projections of yield losses by 2030 in developing world regions.

Uncertainties from four factors were considered in the study: projected temperature change from climate models, projected precipitation change, sensitivity of crops to warming (estimated using time series analysis), and sensitivity of crops to rainfall. In most regions uncertainties related to rainfall were surprisingly small relative to temperature – surprising because year-to-year rainfall variations can be so important to crop production. However, temperature trends are much larger relative to historical variability than rainfall, with mean temperature trends typically twice as big as historical standard deviation while rainfall trends were much smaller than historical variability. In particular, that study identified crop sensitivity to temperature as a key unknown, i.e. there was often a big difference between impacts using a low vs high estimate of temperature sensitivity. Thus, efforts to quantify and reduce impact uncertainties would be well served by a focus on crop temperature responses. Uncertainties in future rainfall trends were less important overall, but emerged as the critical factor in a few key crops, such as rice and millets in South Asia.

10.5 Summary

This chapter has provided a glimpse into results from global and regional assessments conducted over the past decades. The key points are summarized below:

- Global assessments have generally concluded small changes in global prices for a doubling of CO_2, with gains in developed countries balancing losses in the tropics.

Yet these conclusions have been based on a relatively small number of models, and the sources and magnitudes of uncertainty have not been well quantified.

- Projections for CO_2 concentrations more than double preindustrial levels (280 ppm) suggest price increases and negative impacts for food security.
- Regional assessments in China tend to show negative effects of warming but net positive yield changes when including CO_2 fertilization, though many of these studies have unusually large amounts of modeled fertilization (upwards of 40% for mid-century).
- Projections tend to be negative for India and Sub-Saharan Africa even over the next few decades, indicating that these two regions face relatively large risks of crop yield losses. Given the concentration of malnourished populations in these regions (see Chapter 2), these changes are of great importance to global food security.
- The United States will likely experience downward pressure on maize yields, a C_4 crop with limited CO_2 fertilization. Yield losses from warming will likely be balanced in other crops by CO_2 effects in the next few decades.
- Both regional and global assessments would benefit from more explicit consideration of uncertainties from a variety of sources. A particularly important source of uncertainty, among processes currently represented in models, appears to be temperature sensitivity of crops. Effects of pests, diseases, extreme effects, and ozone represent additional factors that are not currently in most models.
- A relatively costly but invaluable approach to quantifying uncertainties is to have multiple modeling groups perform identical experiments with different models. An alternative is to approximate uncertainty in individual model components with statistical distributions, which lends itself to rapid propagation of errors using Monte Carlo techniques.

References

Adams RM, Rosenzweig C, Peart RM, Ritchie JT, McCarl BA, Glyer JD, Curry RB, Jones JW, Boote KJ, Allen LH (1990) Global climate change and United-States agriculture. Nature 345(6272):219–224

Aggarwal PK, Mall RK (2002) Climate change and rice yields in diverse agro-environments of India. II. Effect of uncertainties in scenarios and crop models on impact assessment. Clim Change 52(3):331–343

Boko M, Niang I, Nyong A, Vogel C, Githeko A, Medany A, Osman-Elasha B, Tabo R, Yanda P (2007) Africa. In: Parry OFC ML, Palutikof JP, van der Linden PJ, Hanson CE (eds) Climate change 2007: impacts, adaptation and vulnerability contribution of working group II to the fourth assessment report of the intergovernmental panel on climate change. Cambridge University Press, Cambridge UK, pp 433–467

CCSP (2008) The effects of climate change on agriculture, land resources, water resources, and biodiversity in the United States. A Report by the U.S. Climate Change Science Program and the Subcommittee on Global Change Research. U.S. Department of Agriculture, Washington, DC, USA

Challinor A, Wheeler T, Garforth C, Craufurd P, Kassam A (2007) Assessing the vulnerability of food crop systems in Africa to climate change. Clim Change 83(3):381–399

Darwin R (2004) Effects of greenhouse gas emissions on world agriculture, food consumption, and economic welfare. Clim Change 66(1–2):191–238

Diffenbaugh NS, Krupke CH, White MA, Alexander CE (2008) Global warming presents new challenges for maize pest management. Environ Res Lett 3(4):044007

Dinar A, Mendelsohn R, Evenson R, Parikh J, Sanghi A, Kumar K, McKinsey J, Lonergan S (1998) Measuring the impact of climate change on Indian agriculture. World Bank, Washington, DC

Easterling W, Aggarwal P, Batima P, Brander K, Erda L, Howden M, Kirilenko A, Morton J, Soussana JF, Schmidhuber J, Tubiello F (2007) Chapter 5: food, fibre, and forest products. In: M.L. Parry, O.F. Canziani, J.P. Palutikof, P.J. van der Linden and C.E. Hanson (eds). Climate change 2007: impacts, adaptation and vulnerability contribution of working group II to the fourth assessment report of the intergovernmental panel on climate change. Cambridge University Press, Cambridge, United Kingdom and New York, NY, USA

FAO (2009) Food and agriculture organization of the united nations (FAO), FAO statistical databases. http://faostat.fao.org

Fischer G, van Velthuizen HT, Shah MM, Nachtergaele FO (2002). Global agro-ecological assessment for agriculture in the 21st century: methodology and results. International Institute for Applied Systems Analysis, Food and Agriculture Organization of the United Nations, Laxenburg, Austria

Fischer G, Shah M, N. Tubiello F, van Velhuizen H (2005) Socio-economic and climate change impacts on agriculture: an integrated assessment, 1990–2080. Philos Trans: Biol Sci 360(1463):2067–2083

Huke RE, Huke EH (1997) Rice area by type of culture: south, southeast, and east Asia: a revised and updated data base. International Rice Research Institute, Los Banos, Philippines

Jones PG, Thornton PK (2003) The potential impacts of climate change on maize production in Africa and LatinAmerica in 2055. Global Environ Change-Hum Policy Dimens 13(1):51–59

Kane S, Reilly J, Tobey J (1992) An empirical study of the economic effects of climate change on world agriculture. Clim Change V21(1):17–35

Krishnan P, Swain DK, Chandra Bhaskar B, Nayak SK, Dash RN (2007) Impact of elevated CO_2 and temperature on rice yield and methods of adaptation as evaluated by crop simulation studies. Agric Ecosyst Environ 122(2):233–242

Kurukulasuriya P, Mendelsohn R, Hassan R, Benhin J, Deressa T, Diop M, Eid HM, Fosu KY, Gbetibouo G, Jain S (2006) Will African Agriculture Survive Climate Change? World Bank Econ Rev 20(3):367

Lal M, Singh KK, Rathore LS, Srinivasan G, Saseendran SA (1998) Vulnerability of rice and wheat yields in NW India to future changes in climate. Agric For Meteorol 89(2):101–114

Lin E, Xiong W, Ju H, Xu Y, Li Y, Bai L, Xie L (2005) Climate change impacts on crop yield and quality with CO2 fertilization in China. Philos Trans Biol Sci 360(1463):2149–2154

Lobell DB, Burke MB (2008) Why are agricultural impacts of climate change so uncertain? The importance of temperature relative to precipitation. Environ Res Lett 3(3):034007

Lobell DB, Burke MB, Tebaldi C, Mastrandrea MD, Falcon WP, Naylor RL (2008) Prioritizing climate change adaptation needs for food security in 2030. Science 319(5863):607–610

Matthews RB (1995) Modeling the impact of climate change on rice production in Asia. CABI Los Baños, Philippines

Maurer EP, Duffy PB (2005) Uncertainty in projections of streamflow changes due to climate change in California. Geophys Res Lett 32(3):L03704. doi:10.1029/2004GL021462

Parry M, Rosenzweig C, Iglesias A, Fischer G, Livermore M (1999) Climate change and world food security: a new assessment. Global Environ Change-Hum Policy Dimens 9:S51–S67

Parry ML, Rosenzweig C, Iglesias A, Livermore M, Fischer G (2004) Effects of climate change on global food production under SRES emissions and socio-economic scenarios. Global Environ Change 14(1):53–67

Reilly J, Hohmann N, Kane S (1994) Climate change and agricultural trade. Global Environ Change 4(1):24–36

Rosegrant MW, Cai X, Cline SA (2002) World water and food to 2025: dealing with scarcity. International Food Policy Research Institute, Washington, DC

Rosenzweig C, Iglesias A (eds) (1994) Implications of climate change for international agriculture: crop modeling study. United States Environmental Protection Agency, Washington, DC

Rosenzweig C, Parry ML (1994) Potential impact of climate-change on world food-supply. Nature 367(6459):133–138

Saseendran SA, Singh KK, Rathore LS, Singh SV, Sinha SK (2000) Effects of climate change on rice production in the tropical humid climate of Kerala, India. Clim Change 44(4):495–514

Schlenker W, Roberts MJ (2008) Estimating the impact of climate change on crop yields: the importance of nonlinear temperature effects. NBER Working Paper 13799

Schlenker W and Lobell DB (2009) Robust and potentially severe impacts of climate change on African agriculture. Environmental Research Letters: in review

Tao F, Hayashi Y, Zhang Z, Sakamoto T, Yokozawa M (2008) Global warming, rice production, and water use in China: developing a probabilistic assessment. Agric For Meteorol 148(1):94–110

Tebaldi C and Lobell DB (2008) Towards probabilistic projections of climate change impacts on global crop yields. Geophysical Research Letters 35: L08705, doi:10.1029/2008GL033423

Thomson AM, Brown RA, Rosenberg NJ, Srinivasan R, Izaurralde RC (2005a) Climate change impacts for the conterminous USA: an integrated assessment: Part 4: water resources. Clim Change 69(1):67–88

Thomson AM, Rosenberg NJ, Izaurralde RC, Brown RA (2005b) Climate change impacts for the conterminous USA: an integrated assessment: Part 5. Irrigated agriculture and national grain crop production. Clim Change 69(1):89–105

Tubiello FN, Amthor JS, Boote KJ, Donatelli M, Easterling W, Fischer G, Gifford RM, Howden M, Reilly J, Rosenzweig C (2007) Crop response to elevated CO_2 and world food supply: a comment on "Food for Thought..." by Long et al., Science 312:1918–1921, 2006. Eur J Agron 26(3):215–223

Wang J, Mendelsohn RO, Dinar A, Huang J, Rozelle S, Zhang L (2008) Can china continue feeding itself? the impact of climate change on agriculture. World Bank Policy Research Working Paper Series, number 4470

Xiong W, Lin E, Ju H, Xu Y (2007) Climate change and critical thresholds in China's food security. Clim Change 81(2):205–221

Chapter 11
Where Do We Go from Here?

David Lobell and Marshall Burke

Abstract Some suggestions for future research on food availability, access, and utilization impacts of climate change are presented. Top priorities include better characterization of uncertainties in climate and crop responses, examining income responses to yield changes, and quantifying links between incomes, health, and food security. Many of these questions will require a more interdisciplinary approach than has been typical of past research.

11.1 Introduction

The concepts and models described in this book have evolved out of the work of thousands of researchers over many decades. We collectively know an impressive amount about climate and food systems, but there is far more that remains to be uncovered. In this chapter, we provide our view of the most pressing research needs, aimed to stimulate thought and activity among students and researchers. We return for this discussion to the three factors that comprise food security, as outlined in Chapter 2, and identify several questions for each topic.

Most of the questions below will require insight from multiple disciplines, including climate science, agronomy, crop and animal breeding, ecology, economics, nutrition, and human health. The fact that so many of these questions remain to be explored, even at a superficial level, is a testament to how difficult it can be to work across disciplinary boundaries. Though there are some valid reasons for focusing on single disciplines, many of the obstacles to interdisciplinary work relate to traditional incentive structures for researchers that should be reformed. Only by dealing with the complex reality of food security and by gaining insight from various perspectives can we hope to make big leaps in our understanding of how best to prepare for a warmer world.

D. Lobell and M. Burke
Stanford Univeristy, CA, USA

D. Lobell and M. Burke (eds.), *Climate Change and Food Security*,
Advances in Global Change Research 37, DOI 10.1007/978-90-481-2953-9_11,

11.2 Food Availability

1. Efforts to better quantify and reduce uncertainties related to processes already represented in crop models are perhaps the most critical need for anticipating effects on food availability. A primary means for achieving this will be more experimental studies that manipulate temperature, CO_2, soil moisture, and ozone, both separately and in combination, and for a range of crops of importance to the food insecure. The experiments will be particularly useful in tropical systems where they have been essentially non-existent in the past. A second important approach will be to continue to test existing ecophysiological models with observations of yield and weather variations, at scales ranging from individual fields to entire regions. These same observations should also be used in time-series and panel-based analyses.
2. More effort is also needed to quantify and reduce uncertainties in future climate. No longer should studies use output from just one or two climate models, as projections of 20+ GCMs are now commonly available. Improvements in downscaling techniques will also be of use, although in our view the importance of such efforts is often overestimated, given that downscaling is most critical for rainfall whereas the consequences of rainfall trends will be relatively small compared to warming in many places.
3. Much work is needed to assess the reliability of water resources in irrigated areas, a factor that is just beginning to be considered in tandem with direct yield effects. Similarly, the effects of higher ozone, weeds, pests and pathogens, and sea level rise on crop productivity remain largely unknown and deserving of more quantitative scrutiny. Adding complexity to models should not, in and of itself, be a goal of future research, and in many cases it is unlikely that these factors will significantly change results. Instead, we should seek to identify the domain over which current models are inadequate, and identify not only the key factors needed in those instances but also the size of that domain. Too often a model that fails (or succeeds) in modeling one particular location and set of conditions is dismissed (or applied) in all other situations.
4. There are numerous questions about how fast and effective adaptation measures will be. At the farm level, these include how well farmers' can perceive climate trends amidst substantial variability, how well they understand the response of their crops to these trends, how quickly they can learn and implement new technologies, and what are the risks and likelihood of success for these adaptations. All of these should be amenable to some evaluation, particularly using data from the most recent decade in regions that have been warming most rapidly. There are also important questions about the scope for technology development, and more communication between the crop modeling and crop improvement communities appears particularly worthwhile. Yet often the impacts of future technologies are difficult to anticipate in advance (if we knew what the technologies would be, we'd already have them!)
5. Surprisingly little has been done to evaluate the true scope of cropland expansion in colder regions. All trade models include some component of cropland expansion,

and often this is a critical relief valve for price pressures. But how suitable will the soils in these zones really be for crop production, and how quickly will expansion take place? Again, evidence from recent decades, such as the expansion of wheat in Northern China, should help to better understand these dynamics.

11.3 Food Access

1. As the bulk of poor and hungry populations are in rural areas and have close ties to food production, their incomes could be significantly impacted by both local and global scale yield impacts. Many economic assessments to date have considered GDP growth as independent of agricultural impacts, but this is clearly not the case, particularly in the poorest of countries. Future work should more explicitly consider effects on income and resulting impacts on food security. Critical questions in this area will be the degree to which malnourishment remains concentrated in rural areas, the net position (buyers or sellers) of the poor for key crops, the percent of expenditures on food, and the wage impacts of changes in crop prices.
2. A related question is whether getting out of agriculture represents a viable adaptation strategy for food insecure populations. Clearly some degree of income diversification has been a useful strategy for coping with inter-annual variability. But will it be possible to create enough economic growth outside of agriculture-related industries to reduce the dependence of most people on agricultural productivity? Would governments be wise to promote investment in these other sectors over agriculture in areas that face the most severe impacts?
3. Finally, we lack basic knowledge of how climate change might interact with more traditional development strategies aimed at improving smallholder productivity and incomes. For instance, if adoption of fertilizer and improved agricultural technology is seen as central to improving rural livelihoods, and a primary explanation for current low adoption rates is farmer risk avoidance in the face of a variable climate, then will future changes in climate further inhibit technology adoption?

11.4 Food Utilization

1. Despite recent evidence that higher atmospheric CO_2 will tend to lower protein and micronutrient concentrations (see Chapter 7), there is much we don't know about the eventual health consequences of these changes. Will they be less or more important than associated changes in calorie consumption, and how do the two interact? What management options exist to minimize the reductions in protein or micronutrient levels in crops?
2. Climate change will very likely influence the exposure and infection rates for various human diseases, as discussed in Chapter 2. How will these changes

aggravate, or be aggravated by, changes in food security? What are the critical links between disease and hunger and what are the best points of intervention to improve outcomes?

11.5 Final Thoughts

In pursuing the above scientific questions or the many others we have undoubtedly ignored, and in presenting and communicating the results, we should never lose sight of how difficult it is to predict the future. Models will always be simplifications of reality, and predictions should always be treated with humility and caution. But as discussed in Chapter 1, models provide valuable insight by synthesizing our knowledge of the world and translating it into probabilities of outcomes we care about, and thus help separate the very plausible from the very unlikely. Even if humans tend to err on the side of arrogance more than humility, the danger of understating things we know well is no less than that of overstating those we don't.

In this vein, one thing appears almost certainly true in the twenty-first century: if agriculture and food security are to thrive, they will have to do so in a constantly warming world. The level of climate stability that has been experienced since the dawn of agriculture is a thing of the past; the future will be one of constant change. This need not spell disaster for food security, but we would be wise not to underestimate the enormity of the challenge at hand.

ERRATUM

Chapter 1
Introduction

David Lobell and Marshall Burke

1.1 Why Read This Book?

Erratum to: DOI 10.1007/978-90-481-2953-9_1

2nd Indented line,
4th line should read:

Theory alone cannot refute either of these extreme positions, as there are no obvious reasons why the pace of climate change caused by human activity should or should not match the pace with which we are able to adapt food production systems.

The online version of the original chapter can be found under
DOI 10.1007/978-90-481-2953-9_1

Index

A

Adaptations, 4–10, 32, 37, 40, 42, 73, 75–77, 79, 96, 97, 104–108, 133–152, 156–170, 178–182, 186–189, 194
Autonomous adaptation, 143, 146, 151

B

Basic linked system (BLS), 179, 180, 188
Bayesian model, 41–43

C

C_3, 66–68, 73, 74, 110, 112–114, 116, 117, 119, 124, 170, 180
C_4, 66–68, 73, 74, 109, 110, 113, 116, 117, 124, 170, 181, 184, 185, 190
CERES, 63, 65, 67–69, 72, 74, 92, 140, 183, 185
China, 5, 87, 122, 168, 182–184, 190
Climate extremes, 41, 47
Climate forcing, 32, 33, 37, 47
Climate projections, 31–54, 95–96
Climate uncertainty, 32, 33, 35–38, 40, 41, 53, 54, 75, 147, 188, 189
Climate variability, 54, 86, 134–136, 139, 145
Cognitive bias, 138
Colinearity, 91, 92, 97, 103
Cotton, 63, 100, 101, 107, 122, 187
Crop breeding, 118, 133, 156, 168–169
Crop development, 8, 74, 90, 146–147, 160
Crop growth, 60, 75, 85, 93, 110, 113, 119
Crop insurance, 148, 150
Crop models, 7–9, 42, 43, 59, 77, 86, 95, 119–121, 139, 142, 178, 179, 183, 184, 187, 194
Crop switching, 100, 107, 133, 141
Cross-sectional analysis, 100–104, 106, 107

D

Diseases, 9, 25, 26, 28, 60, 155, 156, 159–161, 163, 190, 195, 196
Downscaling, 34, 35, 95, 194
Drought tolerance, 163, 169, 170

E

Economic growth, 13, 16, 25, 195
Ecophysiological models, 59–80, 194
Emissions scenario, 36, 38, 40, 181
Entitlements, 21
Evapotranspiration, 49, 62, 70, 73, 169, 184
Ex-ante adaptation, 135, 148
Ex-post adaptation, 135, 148

F

Farmland values, 100–103, 107
First-differences, 88–90
Fixed effects, 103, 104, 106, 107
Food access, 11, 14, 15, 21–26, 136, 195
Food availability, 14, 15, 18–21, 29, 194–195
Food prices, 11, 21–25, 29, 135, 145, 146, 150, 178, 187
Food utilization, 14, 26–28, 195–196
Free-air CO_2 enrichment (FACE), 79, 110–117, 119, 120, 123
Functional forms, 93–94, 107

G

General circulation models (GCMs), 32–45, 47, 48, 50, 51, 60, 66, 194
Genetic diversity, 123, 158, 169, 170
Germplasm, 118, 123, 156, 158, 163–165

Global assessments, 8, 20, 178–181, 188–190
Grain quality, 9, 160–162, 168
Green revolution, 18, 157
Greenhouse gases, 4, 6, 10, 31, 32, 35, 37, 44, 47, 79
Growing degree days, 64
Growing season length, 41, 47–52, 139, 140

H
Harvest index, 69, 117
Health, 4, 9, 14, 26, 28, 29, 50, 118, 135, 160, 164, 193, 195
Heat tolerance, 75, 156, 157, 159, 161, 164, 169, 170
Hedonic model, 101
Heteroskedasticity, 87, 97
Household surveys, 15–17, 22
Hybrids, 71, 72, 87, 168

I
India, 5, 8, 24, 168, 181–185, 190
Infrastructure, 28, 112, 133, 142, 144, 146–151, 180
Intergovernmental panel on climate change (IPCC), 10, 33, 35, 36, 38, 40, 44, 45, 59, 77, 122, 185, 188
Irrigation, 8, 49, 60, 71, 73, 75, 76, 87, 96, 99, 107, 133, 139, 140, 142, 143, 149–151, 167, 180, 183–185, 187, 188

L
Land suitability, 143
Landrace, 157–158, 164

M
Maize, 4, 9, 43, 63, 67, 68, 74, 86–88, 90–92, 94, 95, 100, 105, 106, 110, 116, 118, 122, 140, 141, 147, 155, 156, 159, 163, 165, 168, 169, 177, 180–183, 185–187, 189, 190
Micronutrients, 26–28, 118, 160, 161, 166, 195
Migration, 135, 139, 187
Mineral content, 118, 124
Model ensemble, 37, 40–43
Model intercomparison projects (MIPs), 188, 189
Model validation, 40
Multi environment trials (METs), 156, 157, 163–165, 167, 168

N
Net consumption, 22, 23
Net production, 5, 8, 9, 19, 119, 123, 145, 155, 157, 180

O
Omitted variable bias, 104, 107, 108
Open pollinated, 156, 168
Open top containers (OTC), 110, 111, 114, 116, 121
Ozone, 109–124, 188, 190, 194

P
Panel analysis, 103–105, 106, 108, 186
Phenology, 64, 70, 75, 163, 166, 169
Photosynthesis, 60, 63, 65–71, 78, 79, 110, 112, 113, 116, 119
Planned adaptation, 134, 146–151
Plant physiology, 59, 78, 112–113
Planting dates, 8, 60, 71, 72, 74, 76, 79, 139–140, 142, 151, 167, 180, 181
Protein content, 27, 117, 118, 161, 162

Q
Quantitative trait loci (QTL), 164, 169

R
Radiation use efficiency (RUE), 63, 67, 68
Random effects model, 104, 107
Regression, 42, 78, 87–89, 91, 93–95, 100–104, 106–108
Regression bias, 94–95
Respiration, 60, 63, 66, 70, 113
Ricardian model, 101
Risks, 9, 28, 41, 76, 91, 92, 110, 118, 119, 135, 136, 141, 142, 147, 148, 151, 158, 178, 187, 188, 190, 194, 195
Rubisco, 78, 112, 113, 121

S
Selection, 64, 74, 76, 77–78, 90–93, 123, 156, 157, 159, 161–169
Signal detection, 136–138
Soybeans, 63, 74, 78, 107, 110–118, 122, 141, 159, 182, 187
Spatial scales, 5, 63, 188
Stationarity, 96
Statistical models, 32, 41–43, 86, 92, 95, 189

Stomatal conductance, 79, 112, 113, 119, 121, 158
Storage, 69, 71, 73, 104, 105
Sub-Saharan Africa (SSA), 16, 149, 150, 185–187, 190

T

Target population of environments (TPE), 156–158, 162–168
Time series, 35, 48, 85–97, 99, 106–108, 184–186, 189, 194
Trade, 8, 15, 20, 21, 25, 145, 177–181, 187–189, 194

U

Uncertainty, 32, 33, 35–38, 40–43, 50, 52–54, 75, 97, 117, 147, 188–190
Undernourishment, 14–16, 26

W

Water-use efficiency, 110, 156, 160, 169, 170
Weather, 3, 27, 34, 35, 37, 40, 53, 54, 60, 65, 69, 71–73, 76, 77, 80, 85, 86, 90, 93, 94, 96, 97, 99–108, 121, 134, 135, 137, 138, 148, 156, 160, 186, 194
Wheat, 4, 5, 24, 63, 65, 68, 71, 74, 77, 79, 91, 92, 110, 113–123, 138, 139, 141, 149, 155–170, 177, 179, 180, 182–185, 187, 189, 195

Y

Yield trends, 19, 88, 89

CPSIA information can be obtained at www.ICGtesting.com
230311LV00002B/33/P

9 789048 129522